RESOURCE CONSERVATION ECONOMICS AND POLICIES

S. V. CIRIACY-WANTRUP

PROFESSOR OF AGRICULTURAL ECONOMICS
UNIVERSITY OF CALIFORNIA

RESOURCE CONSERVATION

ECONOMICS AND POLICIES

UNIVERSITY OF CALIFORNIA PRESS

BERKELEY AND LOS ANGELES 1952

University of California Press, Berkeley and Los Angeles, California

Cambridge University Press, London, England

Designed by John B. Goetz

CONTENTS

I ORIENTATION OF THE STUDY

1 EMERGENCE OF THE PROBLEM 3

Two important experiences — Changes of technology in renewable resources — Changes of technology in nonrenewable resources — Changes of population growth — Special conditions during the Western explosion — Geographic expansion and resource depletion — Relative size and speed of the explosions — The Western population explosion and political power — Cornucopian complacency or public awareness? — Changes of social institutions — The capitalistic revolution in social institutions — Results of the three changes — Emergence of the problem — Contradictory explanations and solutions — The challenge — Technology of conservation — Economics of conservation.

2 OBJECTIVES OF THIS STUDY 19

Consistent terminology — Economic analysis — Development of public conservation policy — Some limitations.

3 MEANING AND CLASSIFICATION OF RESOURCES 27

Possibilities of misunderstanding — Resources and planning agents — Resources and the social sciences — Focus on natural resources — Resources *ex ante* and *ex post* — Resource use and time — The planning interval — Planning intervals and planning period — Exhaustibility in relation to revenues and costs — Factors affecting changes of revenues and costs — A different classification — Stock resources — Classes of stock resources — Flow resources — The critical zone — Flow resources without a critical zone — Table of resource classes — Primary and secondary resources — Relations between resources — Significant implications.

4 MEANING OF CONSERVATION 48

Conservation does not mean nonuse — Is conservation "wise use"? — Is conservation a static state of use? — A dynamic interpretation of conservation — Quantitative measurement of conservation — State of conservation — Conservation *ex post* and *ex ante* — Measuring not affected by the money veil — Measuring conservation of a composite of resources — Accomplishing conservation — Conservation and "best knowledge" — Conservation and "waste" — Conservation and investment — A maximum of cumulative or sustained use — Maximum sustained use as an economic objective?

II THE PRIVATE ECONOMICS OF CONSERVATION: THE OPTIMUM STATE OF CONSERVATION

5 CONSERVATION DECISIONS 65

Private economics considered first — Conservation decisions and production planning — Relations between use rates defined — Relations through revenues depend on market form — Competitiveness in costs — Complementarity in costs: sunk costs — Complementarity in costs: opportunity costs — Illustrations from agriculture — Illustrations from forestry — llustrations from mining — Illustrations from the oil industry.

6 THE OPTIMUM STATE OF CONSERVATION 76

The optimum state *ex ante* and *ex post* — The optimum state of conservation and maximization — Theoretical determination of the optimum state — Theoretical determination and practical approximation — Approximation step by step — Further simplification — Conservation practices — Objections and limitations — Irrationality — Extramarket values — Uncertainty — Habit patterns — Optimum state of conservation as *ex post* concept — *Ex post* appraisal of habit patterns — Habit patterns and economic calculation — Optimum state of conservation and economic forces.

III THE PRIVATE ECONOMICS OF CONSERVATION: ECONOMIC FORCES

7 INTEREST, TIME PREFERENCE, AND INCOME 97

Why are we interested in economic forces? — General effects of interest changes on conservation

— Interest and the end of the planning period — Do planning agents use market interest rates? — Imperfect markets for assets — The individual time-preference rate — Time-preference rates in relation to income levels — Regression, progression, and fixity — Effects accentuated in underdeveloped countries — Implications for public conservation policies.

8 UNCERTAINTY 111

Uncertainty and uncertainty preference — Allowance for uncertainty — Uncertainty in relation to time — Uncertainty in relation to conservation decisions — Uncertainty discount and conservation — Flexibility and conservation — Hedging and conservation — Pooling and conservation — Spreading and conservation — Shifting of uncertainty bearing to the public — Uncertainty and private capitalism — Uncertainty and the end of the planning period — Uncertainty and habit patterns — Habit patterns and conservation.

9 PRICES AND PRICE SUPPORTS 129

Difficulties in economic analysis — Simplifying assumptions — Effects of different time distributions of price changes — Price changes of "conserving" and "depleting" services — Price changes of "conserving" and "depleting" products — Penalties for "depleting" products — Price supports and conservation policy — Price supports versus income supports — Effects of price changes: conclusion.

10 PROPERTY 139

Economic forces and social institutions — Concept and structure of social institutions — Property as a primary economic institution — Indefiniteness of property rights — Remedies for indefiniteness of property rights — Instability of property rights — Instability versus rigidity of property rights — Imbalance of property rights — Remedies for imbalance of property rights.

11 TENANCY 150

Meaning of tenancy — Effects of tenancy upon conservation decisions — Instability of tenancy — Incidence on owner and tenant — Fixed regressive rents — Imperfections in the markets for assets — The government as landlord.

12 CREDIT 161

Credit in relation to tenure — Credit and uncertainty of tenure — Credit and fixed regressive charges — Credit as an important aid to conservation — Imperfections in the credit system — Conservation clauses in credit agreements.

13 TAXATION 168

Significance of the tax system in resource conservation — The necessity of simplifying assumptions — Variations of taxes over time — Progression and regression in taxation — Tax-saving certificates — Preceding effects exist regardless of types of taxes — Income and profit taxes — Segregation of current costs from investment — Depletion allowances — Taxes on the value of property — Deliberate use of depleting effects of property taxes — Depleting effects of inequalities in assessment — Deferred reassessment of property taxes — Death taxes — Yield taxes — Lump-sum taxes.

14 MARKET FORM 188

Market form and conservation — Monopolistic conditions and rates of use in simultaneous economics — Monopolistic conditions and rates of use in time economics — Monopolistic conditions and fugitive resources — Monopolistic conditions and conservation policy.

15 ECONOMIC INSTABILITY 199

Two aspects of the relations between economic instability and resource use — The meaning of economic instability — Four major effects upon the state of conservation — Adjustment to economic instability in agriculture — Adjustment in instantaneous economics — Adjustment in time economics — Consequences for soil conservation — Adjustment to economic instability in mining — Possibilities for an anticyclical conservation policy — Timing, volume, and flexibility of investments — Coördination between private and public investments — Labor-intensive and material-intensive investments — Regional distribution of investments — Investments and the balance of payments — Mitigating the effects of economic instability — Domestic policies — International policies.

VI THE SOCIAL ECONOMICS OF CONSERVATION: OBJECTIVES AND CRITERIA OF CONSERVATION POLICY

16 CONSERVATION POLICY AND SOCIAL INSTITUTIONS 223

Definition of conservation policy — Formulation and execution of conservation policy — Differences between a social institution and a policy — Social institutions as variables for conservation policy.

17 OBJECTIVES AND CRITERIA OF CONSERVATION POLICY 230

The optimum state of conservation in social economics — Social revenue and cost functions —

Maximizing social net revenues in instantaneous economics — Maximizing social net revenues in time economics — Social revenues and costs in relation to market form — Allocation and incidence of social revenues and costs — Allocation and incidence in conservation policy — Social versus entrepreneurial revenues and costs — Extramarket values and conservation policy — Administrative valuation of extramarket goods — Valuation of collective extramarket goods — Objections to this valuation procedure — Weighting of individual preferences — Postulates concerning social weighting — Distortions of the optimum state of conservation — Compensating conservation policy — The problem of consumer sovereignty — Practical approximations of the social optimum — The social optimum and uncertainty.

18 A SAFE MINIMUM STANDARD AS AN OBJECTIVE OF CONSERVATION POLICY 251

The social risk of irreversibility — The safe minimum standard of conservation — Costs of maintaining a safe minimum standard — The critical zone — Practical definition of a safe minimum standard — Advantages and disadvantages — A safe minimum standard and private enterprise — The safe minimum standard as an economic base level — Safe minimum standard and resource relation — Competitiveness in demand — Competitiveness in supply — Conservation policy if the critical zone has been passed — Safe minimum standard and natural changes of flow.

V THE SOCIAL ECONOMICS OF CONSERVATION: IMPLEMENTATION OF CONSERVATION POLICY

19 DOMESTIC TOOLS OF CONSERVATION POLICY 271

Objectives only one aspect of policy — Direct and indirect tools of conservation policy — Education — Education of the voting public — Education of resource users — Difficulties of early diagnosis — Education of different age classes — Educational agencies — Education is not a cure-all — Zoning — Economic characteristics of zoning — Economic and legal limits of zoning — Compensation in connection with zoning — Zoning and eminent domain — Regulation of practices — Regulation of water use — Regulation of agricultural use of the soil — Regulation of forest use — Regulation of grazing use — Regulation of wildlife use — Regulations of oil and gas production — Regulation of mineral production — Regulation of the processing of resources — Economic and legal limits of regulation — Compensation in connection with regulation — Public management of resources — Direct tools and the safe minimum standard.

20 INTERNATIONAL TOOLS OF CONSERVATION POLICY 304

An international conservation policy? — Four tools of international coöperation — International coöperation in fugitive resources — Conferences and conventions on marine resources — Conservation of Pacific halibut and salmon fisheries — Conferences and conventions on migratory birds — Conferences and conventions on wildlife in general — Needs for international conservation of fugitive resources — International coöperation of broader scope — The first international conservation conference — Pan-American conservation of flow resources — Conservation of flow resources through world organizations — Conservation of stock resources through world organizations — Potentialities of international commodity agreements — Experience with international commodity agreements — Timber — Rubber — Sugar — Stock resources — Commodity agreements and conservation: conclusion — International interests of a U. S. conservation policy — Exploitation of resource monopolies — Accessible and inaccessible resources — Freezing and depletion of domestic resources — Stockpiling of accessible resources — Inaccessible resources — Development of foreign resources — Foreign resource development and population growth — Population increase and conservation of resources — Educational and institutional changes.

21 COÖRDINATION OF CONSERVATION POLICY 347

Limitation of our objectives — Major requirements for coördination — The requirement for over-all coördination — Conservation policy and interrelations of resources — Consolidation within two major spheres — The constitutional framework of conservation policy — Conclusions: directions for improving coördination — A federal planning and reviewing agency — A nonadministrative planning commission? — Self-coördination by federal agencies? — A planning and reviewing agency in state governments — Consolidation of federal executive activities — Political obstacles to consolidation — The Hoover Commission's proposal — Consolidation and the Department of Agriculture — A department of mineral resources and power — Consolidation of state executive activities — Coördination in the federal legislative branch — Consolidation of congressional committees — Professional staffing of congressional committees — Duplication objectionable? — Coördination in state legislative branches — Federal–state coördination — Attempts to provide better federal–state coördination — The Pacific Coast Board of Intergovernmental Relations — Regional organization of federal activities — Size of federal regions — An illustration — Nonconstancy of regional relations — Conclusions concerning the size of federal regions — Interstate coördination — State compacts — Unified river-basin authorities.

APPENDIX

MATHEMATICAL NOTE ON THE THEORY OF CONSERVATION 379

1. Definition of conservation — 2. The optimum state of conservation — 3. Relations between use rates and between resources.

INDEX 383

PART I

ORIENTATION OF THE STUDY

CHAPTER 1

EMERGENCE OF THE PROBLEM

Two Important Experiences

Two experiences of modern Western culture have awakened a widespread and lively interest in conservation. Both may be termed problems of human ecology because they emerge from the interreaction, over time, between culture form and environment.

The first experience teaches that man, as one of the most effective geomorphologic agents, may destroy important and potentially *renewable* sources of his own livelihood and enjoyment. Examples are land destruction through agricultural practices unsuited to the climate, slope, and soil; extinction of animal and plant species through hunting, fishing, and disturbance of habitat; prevention of forest regeneration through certain cutting practices and through periodic burning; replacement of valuable plant associations on ranges through inappropriate degree or timing of grazing; spoliation of scenic and other aesthetic values through road construction, roadside development, and placer mining; interference with the utilization of surface and ground water by denudation of watersheds, pollution, excessive pumping, and damage to infiltration areas.

The second experience brings the realization that *nonrenewable* resources on which modern man depends, especially fuels and metals, are being depleted at a rate which is fast increasing absolutely and sometimes relatively—that is, in relation to the increase of proved deposits through more thorough exploration and new discoveries. In the opinion of many investigators, some of these resources will soon be exhausted, at least within certain countries, or will be obtainable in the future only at much greater costs or more serious strategic risks. It is feared that such exhaustion will threaten the very existence of industrial nations.

Neither of these experiences are new. History shows examples, especially under arid and semiarid climates, in which land, fauna, and flora were destroyed by man, who thereby undermined his own existence.[1] Likewise the exhaustion of nonrenewable resources, such as copper, iron, gold, and silver deposits, has sometimes led to far-reaching changes in the economic and political development of countries or parts of countries. The ghost towns of old mining districts in the Appalachians and the Rocky Mountains have had many predecessors in the Old World since very early days.

Three new factors, however, make the modern experience more striking and create a demand for economic analysis and for a study of remedial public policies. A sketch of these three factors may help in understanding the emergence of the problem in its modern form. These factors are changes of technology, changes of population growth, and changes of social institutions.

Changes of Technology in Renewable Resources

During the nineteenth century, man's effectiveness in changing his environment for better or worse increased manyfold. Resulting conservation problems became especially acute for renewable resources. Only a few illustrations from the United States experience need be cited at this point.

The protective grass sod of the prairies and the high plains

---

[1] One of the earliest discussions in American literature is George P. Marsh, *Man and Nature; or, Physical Geography as Modified by Human Action* (New York: Charles Scribner and Co., 1864; rev. ed., 1871), 577 pp.

could not have been destroyed so rapidly without the steel plow and disk; nor would this destruction have been economical without improved land and sea transportation for wheat to European markets.

The donkey engine and the railroad not only hastened the disappearance of virgin timber, but hampered forest regeneration because they made selective logging uneconomical, destroyed young growth, and left large fire hazards.

Improvements of the centrifugal pump and introduction of the turbine pump made it possible to deplete quickly ground-water reservoirs—the replenishment of which may take decades—and to keep pumping costs from rising in spite of increasing lift.

The passenger pigeon and other game species would still be plentiful if firearms had remained on the technological level of the seventeenth and the eighteenth centuries.

International conventions to protect whales in the Arctic and Antarctic oceans might not have become necessary if power-driven ships and modern explosives had not threatened these last two natural sanctuaries for marine resources. More recently the great mother ship, with complete processing and refrigeration facilities, operating for long periods with a fleet of smaller catching vessels in complete independence from shore bases, has put new pressure on many marine resources.

There have been offsetting influences, particularly through progress in the fields of fertilizer chemistry, plant and animal pathology, and genetics. However, at least in the opinion of many observers, modern technology has not done enough to offset the tendency towards greater potential efficiency in the destruction of renewable resources. Smaller power units based on the internal combustion engine have made better silvicultural practices economical (through caterpillar and truck), but have also made plow and disk more effective (through the tractor) in exposing semi-arid grasslands to wind and water erosion. Modern technology has reduced the relative importance of land for agricultural production, but has also greatly increased the relative importance of land for water-resources development and for recreation. Relations between modern technology and recreation are especially ambivalent. Dams have created new lakes for recreational enjoyment, but have also destroyed anadromous fisheries. Better transportation facilities have opened up many new recreational opportunities, but have also destroyed the particular values of wilderness areas.

Changes of Technology in Nonrenewable Resources

In nonrenewable resources, on the other hand, modern technology has been more generally effective in offsetting depletion. Again, a few illustrations from the United States experience may be cited.

Modern drilling, mining, and refining techniques hasten the depletion of a given deposit, but make it economical to utilize deeper and poorer deposits, with increases in average rate of recovery. Not so many years ago a primary copper-ore body of less than 2 per cent could not be mined economically; at present, copper deposits assaying as low as 0.8 per cent are mined by new underground methods (block caving). In the anthracite mines of the United States, the rate of recovery rose from about 40 per cent at the end of the nineteenth century to around 60 per cent at present.

Beneficiation processes (washing, froth process) in ore treatment have made it possible to rework many old mines and tailings. The selective flotation process with various media and reagents, a refinement of the original froth process, turned mineral deposits formerly too complex for exploitation into usable ores. During the Second World War sea water was "mined" as an "ore" of magnesium.

Modern technology also brings about numerous economies in manufacture (electroplating instead of hot-dipping), in use (more efficient combustion engines, stronger and corrosion-resistant steel alloys), in reclamation after use (utilization of metal scrap and industrial waste), and in the possibilities of substitution.

Substitution is of particular interest if renewable resources are substituted for nonrenewable ones (alcohol from vegetable carbohydrates for "mineral" oil, plastics and paper for metals), and if more plentiful nonrenewable resources are substituted for scarcer ones (glass, porcelain, aluminum, magnesium, for copper and other nonferrous metals).

In balance, however, the demand of modern technology for nonrenewable resources has increased the fear of exhaustion in the minds of the public and many investigators. Strategic considerations are largely responsible. The last two great wars have impressed upon governments and citizens alike the vital importance of owning resources and maintaining safe access to them. Further, the technology of modern warfare has called for unprecedentedly great quantities of nonrenewable resources and, relative to peaceful uses, has left less scrap that could be reclaimed economically.

After the First World War, public action to protect nonrenewable resources, especially oil and metals, was strongly demanded in the public press and in the Congress.

The Second World War has meant a much greater drain upon nonrenewable resources than its predecessor. This is particularly true for the United States. Furthermore, the Second World War and its aftermath have vastly increased the scope of the economic and military responsibilities of the United States. To maintain a resource base which will support these responsibilities has become one of the most important aspects of the United States domestic and foreign economic policies.

Changes of Population Growth

The second factor which, at least to many people, poses the problem of conservation in its modern form, is the great increase in the rate of population growth which started in the Western industrial countries around the beginning of the nineteenth century and gradually spread over most of the globe. Today, the first violent phase of this increase—the population "explosion"—has passed for western Europe, the United States, Canada, New Zealand, Australia, and possibly Japan. In eastern Europe, Russia, South America, and Africa, however, this phase is still in full swing; and the great population centers of Southeast Asia have yet to experience the full impact of the change. The causes of this change and its significance for conservation are closely related to the changes of technology just discussed.

Changes of population depend on changes of birth rates and of death rates. Variations of birth rates, caused by wars, epidemics, and changes of social institutions and of standards of living, are common; but they are comparatively gradual and frequently cyclical. Sudden, violent changes of population are generally due to changes of death rates. The recent population explosion, for example, can clearly be traced to a great and rapid reduction of death rates.

This reduction was caused by the "sanitary revolution," [2] one of the most important aspects of the great technological changes during the nineteenth century. The sanitary revolution was particularly effective in reducing infant mortality, and, therefore, quickly affected the reproductive age classes.

Birth rates (per 1,000 women in the reproductive age classes) did not increase. On the contrary, birth rates soon started to decline, but much too slowly to prevent a population explosion. It takes some time for individual parents and whole populations to become aware of the implication of reduced infant mortality, namely that families and populations can survive with fewer births per family. Even after people have become aware of this implication and have acquired appropriate technological information, adaptive action to the new situation is retarded by religion, tradition, public opinion, and intimate habits.

In the technologically most advanced, most urbanized, and economically most prosperous countries (western Europe, the United States, New Zealand, Australia), the voluntary decrease of birth rates has gone far to bring about a novel "non-Malthusian" equilibrium between death rates and birth rates. In the meantime, however, the time lag in the adaptation of birth rates has brought about a great absolute increase of the population. Still, in the industrial countries of the Western world this greater population enjoys higher living standards than did the people living there before the population explosion took place.

[2] In this term we include not only progress in many fields of medicine but also in the technology of pest control and public sanitation (compulsory vaccination, sewage disposal, pure water supply, urban building codes, etc.).

Special Conditions during the Western Explosion

Some economists have pointed to this conclusion of the population explosion in the Western world and have scoffed at the pessimism of those who fear that the population explosion presently in progress in the rest of the world might not come to such a happy ending. They are supported by optimistic technologists in the fields of soil science, fertilizer chemistry, and engineering who contend that the chances are good that technology will repeat in the future the feat already performed in the past.

Extrapolation of past experience into the future is always hazardous. There is no certainty that the non-Malthusian population equilibrium reached in the Western countries is a stable one. Although the recent upsurge of population in the United States and in some European countries after the Second World War is to some extent cyclical, it should be a warning to those who interpret a rather singular historic experience as a "law."

Extrapolation of past experience into the future is questionable scientific procedure if some important special conditions which helped to bring about a non-Malthusian population equilibrium in the past are not carefully analyzed and weighed with respect to their future significance. Let us consider briefly some of these special conditions.

Geographic Expansion and Resource Depletion

One of these is of particular interest for our study: the Western population explosion was able to expand during the nineteenth century into vast, empty or thinly populated spaces with great, cheaply exploitable natural resources (North and South America, South Africa, Australia, New Zealand) and into more densely populated but—in terms of Western technology and institutions—underdeveloped areas (India, Burma, Malaya, Indo-China, Indonesia, the Caribbean, Africa). In the last-mentioned areas the native population was conquered and utilized as cheap labor for colonial exploitation of rich, virgin resources—again in terms of Western technology and capitalistic institutions.

In both thinly populated and underdeveloped areas, renewable resources (soils, forests, grasslands, and wildlife) were depleted with little regard for sustained yield, and nonrenewable ones (metal ores, fuels, precious stones) with little regard for building up permanent processing industries and diversified local skills. The cheap raw materials and very favorable terms of trade resulting from this depletion helped greatly in the industrial development, in raising standards of living, and in conserving resources [3] in the Western industrial countries. However, this was essentially a temporary condition. At present, there are no areas left in which this kind of depletion could be repeated.[4] Furthermore, this depletion has in many regions permanently reduced the capacity to absorb the result of the population explosion now in progress.

Relative Size and Speed of the Explosions

Another special condition concerns the relative size and speed of the two population explosions.

The Western population explosion started with a population which was small as compared with the population involved now. The population of the Western industrial countries at the beginning of the nineteenth century was about 117 million.[5] The population of that part of the world in which the present explosion is occurring was about 1,739 million [6] in 1930.

Some students assume that a decrease of birth rates will now follow that of death rates faster than in the past because of better knowledge and more public assistance in the field of birth control. It is well to remember, however, that the technical means of reduc-

---

[3] To a considerable extent the increased productivity of agricultural soils in western Europe during the nineteenth and twentieth centuries was based on the import of cheap feed from overseas countries. This import was the foundation of an intensive livestock and manure economy (chap. 20).

[4] Some students point out that there are still large areas of the world which could be made productive. This is true. However, these areas require large initial investments, a long period of "gestation," or high current inputs in relation to outputs. In addition, many political and institutional obstacles to their productivity must be overcome.

[5] Population of Belgium, Denmark, England and Wales, Ireland, Scotland, France, Germany, Holland, Italy, Luxembourg, Norway, Portugal, Spain, Sweden, Switzerland, Australia, Canada, New Zealand, and the United States.

[6] Population of the world (2,140 million) minus population of countries listed in the preceding footnote (401 million).

ing death rates are now far more developed, are applied on a bigger scale, and are better organized (through the United Nations) than in the nineteenth century. Further, modern means to reduce death rates operate in areas where death rates are kept high by mass diseases—for example, malaria—which are unimportant in Western countries. These mass diseases are especially susceptible to an attack by modern technology.

In any event, it is safe to assume that the absolute population increase until a (hypothetical) non-Malthusian equilibrium is reached will be much greater and will accrue faster than that experienced by the Western industrial countries during the nineteenth and twentieth centuries. Population experts estimate this increase for the next fifty years at close to one billion people. Nearly all of this increase will be in the world outside of the Western industrial countries.

The Western Population Explosion and Political Power

Another of the special conditions favoring a non-Malthusian conclusion of the population explosion in the Western industrial countries was their political power to speed their own industrialization through protective trade policies and (if they so chose) to impede industrialization in weaker raw-material-producing countries, especially in colonial and semicolonial dependencies.

Today, some of the countries in which the population explosion takes place are politically not free; some others are politically free but are too weak to prevent indirect interference by others; none of them, with the exception of Russia, is strong enough to enforce terms of trade by political means.

Cornucopian Complacency or Public Awareness?

Enough has been said to indicate that the non-Malthusian population equilibrium (the permanent stability of which is by no means certain) reached by the Western industrial countries during the nineteenth century and the first half of the twentieth can be expected to conclude the much greater population explosion presently in progress only if the people and governments of the world become fully aware of the situation and accept its challenge. Cornucopian complacency based on insufficient analysis of the past may become dangerous.

In this respect the many recent pessimistic books and articles on the subject by neo-Malthusians are not without potential benefit.[7] One may disagree (as this writer does) with the analytical approach employed in most of this literature and with many of the remedies proposed. Still, much has been done to make people aware of the social and political dangers of Malthusian checks operating on a scale heretofore unknown.

This awareness may stimulate private and public effort in the field of resource conservation, which is closely related to efforts aimed at a non-Malthusian population equilibrium. It is fairly well established that realization of such an equilibrium becomes more difficult the lower the economic and educational level of a given population—and vice versa. Resource depletion is related in many ways, as a cause as well as an effect, to a low economic and educational level.

Changes of Social Institutions

The third factor which, next to changes of technology and of population growth, has done most to focus attention upon the modern problem of conservation is an institutional one.

Before the nineteenth century in Western culture, as today in many other cultures, the utilization of renewable resources (agricultural land, forests, ranges, wildlife) was directed largely toward home consumption and was closely regulated by social institutions and customs. Utilization for sale at markets was limited by available transportation facilities; for the bulky prod-

[7] Ward Shepard, *Food or Famine: The Challenge of Erosion* (New York: The Macmillan Co., 1945), 225 pp.

William Vogt, *Road to Survival;* with an introduction by Bernard M. Baruch (New York: William Sloane Associates, Inc., 1948), 335 pp.

Fairfield Osborn, *Our Plundered Planet* (Boston: Little, Brown and Co., 1948), 217 pp.

U. S. Department of State, *Proceedings of the Inter-American Conference on Conservation of Renewable Natural Resources, Denver, Colorado, September 7–20, 1948* (Washington: U. S. Department of State, 1948), 782 pp. (U. S. Department of State Publication 3382. International Organization and Conference Series II. American Republics 4).

ucts of agriculture and forestry this meant largely transportation by water.

An economy that produces mainly for home consumption utilizes its resources more conservatively, as a rule, than one that produces largely for sale at distant markets. Of these two economies the former is usually more diversified, more immune from price fluctuations, and less subject to adverse effects of taxation, credit, and tenure. Why these characteristics are important for conservation will be explained in detail later.

As shown elsewhere,[8] feudalism, manorialism, the stationary-field-grass system, certain forms of the three-field system, and other social institutions which restricted the freedom of private property influenced decisively the utilization of land, forests, and wildlife in most European countries until the beginning of the nineteenth century. This statement holds even for countries where these institutions had been formally abolished at an earlier date. Their influence favored conservation for reasons explained in the article cited in footnote 8.

The Capitalistic Revolution in Social Institutions

The nineteenth century brought radical changes in the economic and social institutions that regulated the use of renewable resources. Because of the technological revolution in land and sea transportation, production for distant markets became possible. Institutional restrictions on private land use were rapidly relaxed. The profit motive became more important than the motive of maintaining an accustomed level of home consumption. Land became a form of investment, a capital good that could legally be used, depreciated, or maintained in any way the owner found profitable.

Europe still felt the influence of earlier institutions under which land had been, for the "upper" classes, the key and symbol of social distinction and political influence; for the "lower" classes, the assurance of economic security and personal and political freedom.

---

[8] S. V. Ciriacy-Wantrup, "Soil Conservation in European Farm Management," *Journal of Farm Economics*, Vol. 20, No. 1, February 1938, pp. 86–101.

This influence, however, did not exist in the new continents—the Americas, Africa, Australia, and the colonial Far East—into which Western economies expanded with the help of the technological revolution. In these areas and in the "no-man's land" of the high seas, capitalistic exploitation was subject to few institutional restrictions. To be sure, in the United States, as in other countries, capitalistic exploitation continued to be subject to "necessary" and "reasonable" restrictions under the police power. In the past, however, public opinion and the courts have not interpreted these concepts in a way to avoid some serious social consequences of resource depletion.

Social institutions are developed by man and can be changed. In a study on conservation, they may be regarded both as independent and as dependent variables. In other words, they are of interest for our study both in an analysis of the behavior of resource users and as tools of public conservation policy.

Results of the Three Changes

The results of these changes of technology, population growth, and social institutions were striking. The wilderness areas of the world, which were still large in the beginning of the nineteenth century, were quickly conquered. The "primitive" people occupying them were destroyed or pushed back into areas topographically and climatically hazardous for permanent cultivation or were used as cheap labor in plantation agriculture and capitalistic mining. Raw materials in previously unknown quantities and at lower prices helped greatly to raise the living standards of Western industrial countries during a period when their populations were likewise increasing sharply.

Despite this increase of living standards, however, some results alarmed even Western people. After the spring flood of economic exploitation had passed on to new areas, men noted the disappearance or depletion of marine resources, birds, and game. The flood had left behind it accelerated soil erosion, forests reproducing slowly or not at all, deteriorating grasslands, polluted streams, falling ground-water tables, decaying mining towns, and stranded workers.

In the United States the speed and the magnitude of this experience, with the realization that the geographic limits of expansion would soon be reached, gave rise to the conservation movement. This started in the 1870's with the proposal (not realized until 1891) by Secretary of the Interior Carl Schurz that the unreserved public timberlands be set aside as national forests for the protection of watersheds; and with two reports of the American Association for the Advancement of Science (1873 and 1890). The conservation movement reached its first peak shortly before the First World War in the spectacular domestic and international campaigns sponsored by President Theodore Roosevelt and Governor Gifford Pinchot. The reasons for the somewhat meager achievements of these campaigns will be discussed later.

Emergence of the Problem

At that time the modern problem of conservation began to emerge in more articulate form. The following questions were raised: First, why does utilization of potentially renewable resources lead to their drastic and often irreversible impairment, and what factors determine the rate of depletion of nonrenewable resources? Second, how far is the impairment of renewable resources socially undesirable, and what interest has society in the rate of depletion of nonrenewable resources? Third, what changes in social institutions and what kind of public action would implement the interests of society in resource utilization?

Contradictory Explanations and Solutions

The "why" has been answered in various mutually contradictory ways—for example, by pointing to the shortsightedness of individual resources users, to their ignorance, to their ruthlessness in following the profit motive and their shrewdness in circumventing the laws, to the wastefulness of competition, to the greediness of monopoly, to increases in prices, to depressions, and to other "causes." Some of these explanations still appear plausible to the public and many experts alike.

Denying a special concern on the part of society, some students have argued that destructive exploitation was necessitated by economic forces prevailing at the time and has produced a desirable transformation of resources—of primeval forests into agricultural land, homes, and railroads; of cheap virgin lands into food, and raw materials for industry; of rich mines into tools and equipment.

Similarly it has been suggested that progress in technology will solve all resource problems; that "energy being constant," human inventiveness has merely to solve some technical difficulties in the economical transformation of energy; that this solution is a question of need; and that the only resources worth conserving are human intelligence and a cultural climate favorable to technological progress. The possibilities of making atomic energy available have given strong support to this view.

On the contrary, other students have asserted that constancy of energy as a physical principle has little relevance for conservation because the whole may be constant while that portion with which the social sciences are concerned may change; that transformation of energy into forms which could be substituted for important classes of resources is not an economical possibility in the foreseeable future; that society—in contrast to individuals—has the same interest in unborn generations as in the present one; and that a *laissez-faire* policy regarding resources constitutes suicidal negligence.

Those who favor public action represent all shades of opinion. Some advocate complete socialization and public management of resources. Others believe that education—for example, through an agricultural extension service and through demonstrations undertaken by a soil conservation service—will be sufficient. The most pessimistic investigators feel that the problem is unsolvable without bringing resources and human population into better balance through birth control, and even through withholding information which might interfere with the "beneficial" effects of malnutrition, diseases, and epidemics upon death rates.

The Challenge

All these contradictory explanations, arguments, and opinions have been advanced by honest and intelligent people;

and most contain a measure of truth. They are therefore a challenge to understand the factors that affect the distribution of resource use over time, to inquire into the objectives of public policy with respect to this distribution, and to seek tools with which to implement policy objectives. This challenge will be taken up in the succeeding pages.

What is meant by "distribution of resource use over time" will be discussed in the following chapters. The intertemporal distribution of resource use has technological, economic, social, and political aspects of its own. Attention may be focused upon these, rather than upon the related, equally important, but more commonly discussed aspects of distributing use over different human wants with identical time dimension. This special focus is the distinguishing element of a study on conservation.

Technology of Conservation

Most scientific investigations have been directed toward the technology of conservation. This technology is concerned both with the *natural* [9] decrease, increase, deterioration, and amelioration of resources over time and with the similar changes caused by *human action.* For example, soil conservation is studied under soil physics and chemistry, physical geography, and climatology. The conservation of oil, natural gas, and minerals interests geologists, mineralogists, and petroleum and mining engineers. Many biological sciences, especially ecology, nutrition, pathology, and genetics, have built the technological basis for conserving plant and animal resources. In conserving water resources, hydrology, irrigation engineering, and forestry coöperate. Tangible cultural resources, such as buildings, communication systems, machines, and equipment, are the concern of architecture, structural and mechanical engineering, and industrial chemistry. These and other applied sciences, and the more fundamental sciences from which they are derived, have gathered a wealth of information that is useful in analyzing the economic and social problems of conservation.

---

[9] The term "natural" is used in this study in connection with changes that take place without human action or influence.

Economics of Conservation

The economics of conservation attempts to understand the distribution of resource use over time in terms of the relations between technological knowledge, individual motivation, and social institutions; to analyze the economic forces affecting changes in this distribution; and to scrutinize criteria for that distribution that is alleged to be privately or socially "best." The economics of conservation serves, therefore, as a basis for formulating and implementing public policies that aim to protect or to change a given time distribution. Always, however, understanding for its own sake is worth while and is also a prerequisite to prediction and public action. Such an understanding is the main objective of our study.

CHAPTER 2

OBJECTIVES OF THIS STUDY

Consistent Terminology

For the student of semantics the field of conservation offers rich material. The term "conservation" has many denotations and connotations in popular as well as scientific language. Considerable emotional and ethical appeal attaches to some of these meanings. Conservation has become identified with "wise use"—whatever that means to various people—and usually implies a strong positive value judgment.

Naturally enough, a concept not clearly and consistently defined, with strong emotional appeal, whose economic rationale is more difficult to grasp than might appear at first sight, has been utilized for furthering objectives that are not necessarily in harmony with it. Such objectives have been at various times "breaking of monopoly power," "eliminating wastes of competition," "stabilizing prices," "increasing employment," "socialization of resources," "changes in income distribution," "alleviating population pressure," "birth control," "expanding national sovereignty over the high seas," and several others.

All these objectives may be socially justified under certain conditions. However, they should not be confused with conservation; and the conservation argument in favor of public policies to bring these objectives about should not be used without a thorough analysis of the economic and social interrelations involved.

Failure to undertake such an analysis will render conservation policy less effective or will even lead to a policy that is contrary to conservation objectives. Bona fide conservation will thereby become discredited in the eyes of resource users and the public. It will therefore become more difficult to design effective conservation policy and to make it acceptable in a democracy. For such acceptance, education—that is, broadening the basis of understanding and independent judgment—must be relied upon to a considerable degree. Without confidence in the consistency and competence of those who promote policy, neither the resource users nor the public in general can be expected to coöperate.

Our preliminary objective, therefore, is agreement about the meaning of "conservation." The attempt will be made to separate the economic and technological aspects of conservation from ethical, metaphysical, and other connotations. It will appear that the concept "conservation" can become neutral in terms of value judgments (*wertfrei*) and a highly effective tool in economic analysis. This gain, however, entails an unavoidable sacrifice: the concept developed may not be identical with the meaning to which some readers are accustomed. The reader's patience in bearing the necessary terminological prelude of chapters 3 and 4 is, therefore, solicited.

Economic Analysis

The second objective has already been mentioned (chapter 1), namely, the understanding of a given distribution of resource use over time, the study of criteria for an allegedly "best" distribution, and, especially, the analysis of the economic forces affecting intertemporal distributions.

Such an objective is of practical importance for resource users and policy makers, since the most detailed information about the technology of conservation does not provide criteria for distributing resource use over time, nor for the ability of public policy to affect such a distribution.

Aside from its practical importance, economic analysis of conservation is attractive as a scientific inquiry in itself. It affords an opportunity to test certain tools that economic theory is developing for problems connected with time, uncertainty, and social welfare. These problems warrant more attention if theory is to present a systematic view of economic facts, a view that is rounded, clear, and useful as a basis for decisions by individuals and public bodies. Existing economic theory does not satisfy these requirements. For our purposes, three aspects will need special attention.

First, the assumptions, often only implied, regarding the motivation of economic decisions by private individuals and regarding proposed criteria of public policy need to be clarified.

Second, the selection of variables considered by economic theory (that is, the level of abstraction) tends frequently to obliterate relevant relations, especially between variables dependent on time.

Third, existing economic theory is not always suitable for practical simplifications and approximations, which in the actual world are necessitated by the type of information available to the agents who make conservation decisions. In any applied field of science, the task of theory is not ended by explaining relations and by pointing out the solution of any problem in the abstract. Theory should also aid in gathering information and in making the simplifications and approximations that are needed in reality. Practical short cuts, to use an analogy, are safe in difficult territory only if one has a reliable map, an abstract explanation of the relevant topographical features.

The task of economic analysis may be divided into two parts: the private and the social economics of conservation.

In the private economics of conservation let us inquire how individual resource users reach decisions on the distribution of use rates over time. What are their problems? With what validity can one make statements about their motives and objectives? Is there an optimum for them in the intertemporal distribution, and what is it? What major forces in the economic environment affect entrepreneurial decisions and influence the optimum of conservation? Do these economic factors affect conservation even if the motivation of decision-making agents is ignored or misinterpreted or—as an extreme—if the actions of resource users are regarded as random? How can economic theory assist in solving these issues? Can theoretical solutions be so formulated as to aid, in economic reality, the resource users themselves and also those who wish to influence the users' behavior? What are the limitations of theoretical solutions and of realistic approximations? Answers to these questions will be sought in chapters 5 to 15.

In the social economics of conservation let us inquire whether, why, and under what conditions conflicts exist between an intertemporal distribution of use rates brought about by decisions of individual resource users and a distribution that may be regarded as more desirable in the interest of a social group. Is there a social optimum in the intertemporal distribution of use rates, and what is its relation to the private optimum? How far is economic theory useful in defining the social optimum, under what conditions and limitations? Can the social optimum be so formulated through approximations as to be operationally useful for public policy? What are the causes of differences between social and private optima, and how can such causes be attacked through public action? These and other inquiries will yield general criteria and objectives of conservation policy (chapters 16 to 18). Alternative tools of conservation policy by which objectives may be implemented can then be investigated.

Development of Public Conservation Policy

Among these alternative tools, existing social institutions need particular attention. Economic forces affecting conservation decisions of individual resource users are strongly influenced by social institutions. Such influences, if unrecognized, may lead to socially undesirable resource utilization; but, given a knowledge of them, men can utilize social institutions effectively as tools of conservation policy or can modify institutions in order to prevent interference with conservation objectives. Examples for relevant social institutions are those affecting interest, income, uncertainty, prices and market form, the definition of property, the type of tenure, the size of holdings, the tax system, and creditor-debtor relations. In order to avoid repetition, such indirect tools of conservation policy will be discussed when the impact of economic forces upon private conservation decisions is considered (chapters 7 to 15).

Besides indirect public action through modification of existing social institutions, there are numerous alternatives of direct control. Such controls range from mild forms of zoning against certain resource uses to the prohibition or requirement of specific practices and the fixing of production and sales schedules. Compensation may or may not be provided for such interference with private objectives. If controls are rather comprehensive with respect to the actions of individual planning agents, socialization of resources is accomplished in all but name. In that event, outright public management may be preferable. For the domestic field, such direct tools of conservation policy will be taken up in chapter 19.

In the conservation policy of many nations, tools are used that operate in the field of international relations. Such international tools of conservation policy are needed with some renewable resources (high-seas fisheries, migratory birds) that are international in character. They are also needed to cope with problems encountered in military aspects of resources, in foreign investment for resource development, and in repercussions of resource use on human populations—for example, changes in numbers and the push and pull to migrate. Enlightened national self-interest suggests coöperation with other nations in order to mitigate the conflicts that emerge from divergent national objectives in conservation policy. Some lessons can be learned from past attempts of two or more nations to coöperate in applying international tools of conservation policy and from United States participation in such attempts. For resources with which considerations of military security are paramount, international tools of conservation policy are needed that can be used by an individual nation if coöperation with others is not sufficient; such tools must be considered for the United States in the present state of world affairs. The various international tools of conservation policy are discussed in chapter 20.

Tools of conservation policy (both domestic and international) for the same and for different resources are interrelated. Some tools of conservation policy are complementary in their effects; others are independent or competitive. Conservation policy is related in many ways to policies concerned with other objectives —those relating, for example, to full employment, international trade, distribution of income, and national security. Various administrative levels (federal, state, local) participate in the implementation of conservation policy. On any one level several executive agencies are frequently concerned. The legislative mandate for conservation policy appears in many acts, treaties, compacts, and ordinances. The multiplicity of executive activities and of legislative purposes has led to conflicts and waste. For all these reasons, the governmental agencies, both executive and legislative, that deal with conservation policy must be coördinated. Possibilities of such coördination will be reviewed in chapter 21 in the light of recent experience and proposals in the United States.

Some Limitations

A few remarks may be permitted concerning the general scientific character of the approach to the objectives just outlined. Universal historical, as well as geographic, validity could be obtained by choosing an appropriate level of abstraction. On the other hand, if an attempt is made to apply economic theory to the solution of real problems of public policy, the degree of concreteness required necessarily limits validity historically as well as geographically.

To solve this conflict, different levels of abstraction will be used in different portions of the study. For example, the concept of conservation and certain theoretical aspects of the degree of conservation that is the objective of or is actually realized by private individuals and social groups will be treated with the aim of attaining rather broad validity. Conversely, practical approximations to theoretical solutions, and the discussion of tools of conservation policy, are valid primarily for a modern Western economy that is based on exchange and that has an important private sector in the field of resource utilization. In the applied portions, conditions within the United States will receive paramount attention.[1] Restriction to the United States would be undesirable, however, because other countries have had a longer experience with conservation or face a more critical situation.

There is one further reason why different levels of abstraction are desirable. This study attempts to reach various groups of readers with varying training in the methods of economic analysis. These groups include the economic profession itself, government executives, legislators, and the broad and highly important group of intelligent resource users and citizens who realize more and more that conservation of resources is of vital and immediate interest to them and their country. Some basic economic and social issues of conservation will be stated in such a way that they can be understood by the noneconomist; the same holds for the implementation of conservation policy. Many issues, however, need to be subjected to a more rigorous economic analysis. Sections dealing with them are mainly directed to professional economists—who at present regard the field of conservation more as the domain of the idealist, the dilettante, and the politician than as a vital and theoretically interesting subject for social research. The reader with less training in economics may wish to pass over such sections.

This study presents neither a history of conservation nor a detailed blueprint of conservation policy for individual resources in specific countries. Emphasis will be placed on a logical connection between presently available facts through a theory (hypothesis) of cause and effect, on an analysis of the typical problems which arise in the conservation of most resources, and on the principles which apply to policy implementation in most countries. An attempt will be made, however, to bring in constantly the history and cultural geography of conservation as illustrations and as checks on the agreement of theory and observation.

[1] In the Spanish edition of this study an additional part deals more specifically with Latin American countries.

The economics of conservation cannot deal with the totality of the intertemporal aspects of relations between man and his environment. The attitudes, arts, languages, customs, taboos, and religions of many societies from the dawn of history to the present day are interrelated with the conservation complex. The resource base and the causes and effects of its changes have received comparatively little attention from otherwise painstaking students of cultural development. The vacuum cannot be filled here; but perhaps this omission will stimulate contributions by specialists in the sister social sciences of sociology, cultural geography, and history of social institutions.

CHAPTER

3

MEANING AND CLASSIFICATION OF RESOURCES

Possibilities of Misunderstanding

If the concept "conservation" is used in economic analysis, various connotations may be carried over from popular into scientific usage. This danger is common for many economic terms. The social sciences labor under a disadvantage as compared with most natural sciences because their terminology is taken from everyday language and is not created expressly from Greek or Latin roots. Terms like "land," "labor," "capital," "utility," "income," "investment," and "savings" are used by economists with specific meanings, which differ considerably from those intended by the noneconomist.

The possibilities of misunderstanding are especially great in the present discussion. The term "conservation" was deliberately adopted in 1907 as the name for a popular movement in the United States.[1] The historical setting and the driving forces of that movement were mentioned in chapter 1. The term was chosen

[1] Cf. Gifford Pinchot, "How Conservation Began in the United States," *Agricultural History*, Vol. 11, No. 4, October 1937, pp. 255–265, particularly pp. 262 and 263.

to awaken a nation to the fast-growing problems of its resources.

Previously, "conservation" had been used mainly in the sense of keeping social institutions, prerogatives, ideals, and the like, "unimpaired," in the "*status quo*," on the "present level." In relation to resources, terms like "preservation," "protection," and "restoration" appear in the literature during the last quarter of the nineteenth and the beginning of the present century.

Since its adoption, however, the term "conservation" has been associated not only with resources that can be kept unimpaired despite use, but also with those that are necessarily depleted through use. To cite only a few examples, people commonly speak of conserving oil, natural gas, coal, ores, and already produced strategic metals. Before defining the economic meaning of "conservation," therefore, one must consider the classes of resources with which the term may be used.

Resources and Planning Agents

The concept "resource" presupposes that a "planning agent" is appraising the usefulness of his environment for the purpose of obtaining a certain end. Such an appraisal, in turn, presupposes certain technological means at the disposal of the agent and certain institutions (laws, customs, and so on) of the society in which he operates. The planning agent may be a natural person (an individual householder or entrepreneur), a legal person (a corporation, a trade association), or the whole community represented at various levels (federal, state, county, public district) by its government. A resource, therefore, is a highly relative concept changing with the ends-means scheme—that is, with the planning agent, with his objective, with the state of technology, and with existing social institutions.

Resources and the Social Sciences

It follows that resources, their scarcity, their depletion, and their conservation are concepts of the social sciences *par excellence*. Failure to understand this must lead to confusion: the ratio between resources and population is frequently discussed as if resources were a fixed, physically or biologically given quantity in a closed system. Instead, resources are variables in a socially most significant function in which man, his objectives, his knowledge, and his institutions are other variables.

Frequently it is necessary to hold some of these variables constant in the economic analysis of others. For example, the several resources represented by an acre of agricultural land may be held constant in an economic analysis of inputs—that is, of agricultural intensity. However, this expedient for the purpose of partial analysis should not be confused with "constant" in the sense of a physical datum.

The foregoing point needs emphasis in view of the recent spread of neo-Malthusian doctrine. Land is a physically given constant for a deer herd. Here it is sound ecological reasoning to compute "carrying capacities" and to strive for an adjustment of size of herd to these limits. The same reasoning applied to human populations overlooks that land and other resources are interdependent variables in the ends-means scheme. Which are regarded as dependent and which as independent variables, which ones are held constant and which are varied in a partial analysis, is not determined by some fixed physical characteristic of a resource but by the objectives of the study.

Sometimes the differentiation is made between "physical" resources and "economic" resources. The former term is used for those resources which cannot be economically utilized. This confusing terminology may suggest that some resources are not the concern of the social sciences. Nothing could be further from the truth. Resources can thus be termed "physical" only *after* an appraisal on the basis of a given ends-means scheme—in the above sense. Through such an appraisal they may be designated as "submarginal" for certain uses at a certain time and place.

Focus on Natural Resources

According to common usage, one may differentiate three broad classes of resources: natural, cultural, and human. This trinity corresponds to the trinity of "factors" of production—land, capital, and labor—as used commonly in economic texts. The former terminology has this advantage: it is frankly general and is not affected by the many specific connotations that have been attached to "land," "capital," and "labor" in the course of their history.

In principle, natural, cultural, and human resources are of equal interest for economic analysis. Actually their relative importance will differ with ends-means situations. This study, however, deals explicitly only with natural and to a smaller extent with "tangible" cultural resources. The latter are producer and consumer goods that are durable over a period of utilization. There are various reasons for this limitation.

First, the task of dealing properly with the conservation problems of all resources is simply too great for individual research workers. Division of labor is best accomplished on the basis of resource classes which have important characteristics of their own.

Second, sociology, law, and other social sciences are better equipped to deal with the peculiar problems of human and "intangible" cultural resources (for example, the giftedness of a certain population as artisans, artists, and scientists; the Constitution and the Magna Charta) than economics is. Economic analysis, though by no means without significance in these resource classes, would not cover all or even most of the significant aspects.

Third, economic analysis will be largely in terms of pecuniary value. The shortcomings of this approach are relatively less for natural and tangible cultural resources.

All resources are interrelated, in ways that will presently be explained. Unavoidably, therefore, we shall discuss repeatedly and at some length certain aspects of human and of cultural resources. *But the term "resources" will refer mainly to natural resources.*

Resources Ex Ante *and* Ex Post

A resource appraisal by planning agents is concerned with *expected* use. Thus, if resources are considered from the point of view of the planning agent under the technology and social institutions under which he operates, we may say that resources *ex ante* are being considered. This point of view is highly significant for the economic analysis of conservation.

Sometimes, however, it is important to view resource use not as it is expected by planning agents, but as it actually can be observed historically and statistically under the influence of changing technology and institutions. Resources from this point of view may be called resources *ex post.* Generally, the point of view of the analysis is apparent from the connection in which the term "resource" is employed.

Resource Use and Time

Resource use extends over time. The stream of resource use over time in its aggregate will be called "cumulative use." It is measured in appropriate *physical* units, for example, tons, calories, kilowatt-hours, acre-feet.

The concept "cumulative use" must be sharply differentiated from the concept "rate of use." Such a rate may be an average rate—that is, per unit of time over any period of use considered—or it may be an instantaneous rate, that is, the limiting value which the average rate would assume if the period should become very small, or, more correctly, should approach zero.[2] The distinction between a "period" and an "instant" of time is like that between a line and a point: a period has some length or extent, but an instant has none.

This study deals mostly with instantaneous rates of use. In order to avoid cumbersome language they will be called, for short, "use rates" or, simply "use"—if confusion with "cumulative use" is not likely.

The Planning Interval

The concept "rate of use" requires a further explanation in relation to the time interval employed in planning or studying cumulative use—that is, the planning interval or, for short, the "interval."

This "interval" is defined as that period of time within which

[2] In mathematical terms, instantaneous rate of use is the first derivative of cumulative use with respect to time. If we have an instantaneous rate x of a resource X_t then $x_t = \frac{dX}{dt}$.

changes of use rates and of other economic variables can be neglected in the analysis. Always, only one rate of use is considered in an interval.

The actual clock-time extent of an interval may be any extent of time—a day or part of it, a week, a month, a year, or a number of years (perhaps a crop rotation or a cutting cycle)—according to the nature of the economic problem to be solved. For example, in wheat production the interval is determined by the seasons and is, therefore, a year; in hog production the interval is approximately three quarters of a year as determined by the time required for breeding and fattening; in the production of milk and eggs, output may be planned by the month; the pumping from an oil well may be planned by the week; on the other hand, in the production of lumber the planning interval may be several decades.

The concept of a planning interval, as just defined for economic analysis, and the concept of an instant, as defined mathematically, have a characteristic in common which is relevant here: the time considered is one within which changes of use rates may be neglected. In previous publications of this author, therefore, the planning interval was called an "instant." It became apparent, however, that with some readers this terminology led to semantic "blocking" when the planning interval had a considerable extent in terms of clock time, as, for example, in farming and forestry.

Planning Intervals and Planning Period

Planning agents in appraising a resource (resources *ex ante*) or the observer in analyzing resources historically and statistically (resources *ex post*) may adopt any period of time over which use is planned or over which use is considered in the analysis. For short, this period of time will be called the "planning period." Planning intervals are the units of the planning period.

We shall see in chapter 6 that, in *planning* resource use (resources *ex ante*), the extent of this period is, in principle, a variable—both in terms of number of intervals and in terms of clock time—and that this extent is determined simultaneously with the optimum state of conservation. Thus, from the point of view of planning agents, the planning period is always *ex ante:* if experience leads to a change of plans, a new planning period is said to begin at the interval the change of plans is decided upon.

After these precautions against a misunderstanding of concepts used, we can now clarify an aspect of resources which is generally regarded as their most important characteristic for classification —their exhaustibility or inexhaustibility.

Exhaustibility in Relation to Revenues and Costs

Resources are usually classified with respect to their exhaustibility.[3] "Exhaustible" and "inexhaustible" are concepts that are meaningful only if employed in the economic sense. Long before a given resource is physically used up or even appreciably diminished, it may be exhausted in the sense that further utilization is indefinitely discontinued in spite of continuing human wants. The reason is that, in any additional interval, the costs of producing any possible quantity of this resource may be larger than the revenues that could be obtained from this quantity. On the other hand, a resource may be inexhaustible, in the sense that utilization continues indefinitely, even though it is relatively limited in physical quantity, as compared with other resources. The explanation is that it is economically feasible to use only very small quantities of this resource per unit of time.

To define revenues and costs at this point would require an analysis of the economic objectives of planning agents in private and in social economics. This analysis will be undertaken later. Meanwhile, revenues and costs may be regarded either as pecuniary values established by a market (and appearing as positive and negative pecuniary entries in production budgets), or as

---

[3] Bernhard E. Fernow, *Economics of Forestry* (2d ed.; New York: Thomas Y. Crowell and Co., 1902), p. 10.

L. C. Gray, "Economic Possibilities of Conservation," *Quarterly Journal of Economics,* Vol. 27, No. 3, May 1913, pp. 497–519.

Ralph H. Hess, "Conservation and Economic Evolution," in Richard T. Ely, *et al., The Foundations of National Prosperity* (New York: The Macmillan Co., 1917), p. 117.

Erich W. Zimmermann, *World Resources and Industries; a Functional Appraisal of the Availability of Agricultural and Industrial Resources* (New York and London: Harper and Brothers, 1933), p. 796.

"real" values in terms of utilities and disutilities of individual planning agents. All that is necessary to point out here is that we are dealing mainly with changes in revenues and costs over time rather than with changes along revenue and cost curves in instantaneous economics. Because revenues and costs of differently dated rates of use are interrelated (for reasons soon to be discussed), it is advisable to speak of revenue and cost *functions* rather than curves or schedules.

Factors Affecting Changes of Revenues and Costs

All factors that cause changes of revenues and costs over time are, in principle, important for the problem of exhaustibility. Among these factors, obviously, are changes of wants and changes of technology.

However, two additional factors are of special interest in the economics of conservation. These factors operate even if changes of wants and of technology are excluded by assumption. These factors are, first, the effects of cumulative use of resources; second, variations in the physical quantity and quality of resources over time, regardless of use.

Cumulative use of resources may change revenues over time because the accumulation of durable products and scrap affects the demand. For example, accumulation of cut diamonds influences the demand for uncut diamonds; accumulation of metal scrap influences the demand for ores.

Cumulative use also changes costs—and, therefore, the supply—over time for many technological reasons. In mining, eventually, shafts have to be driven deeper, thinner veins followed, ores of lower metal content utilized. Oil and water wells must be drilled deeper; and pumping costs increase because of the greater lift. In agriculture, plant nutrients must be supplied to the soil through fertilizer; or other practices that increase unit costs must be adopted. In fisheries, costs have increased greatly through depletion of fishing grounds.

Changes in quantity and quality of a resource over time, regardless of use, may mean increases or decreases, deterioration or amelioration, continuously or discontinuously, at a constant or at a varying rate. For example, the quantitative and qualitative changes taking place in timber during its period of growth affect both revenues and costs. The blowoff of natural gas affects the costs of both gas and oil. Changes over time in the total quantity of fish and other wildlife because of diseases, natural changes in environment, and the like, may affect revenues and costs decisively. Although changes in quantity and quality of land, regardless of use, are generally slow, they sometimes take place at an economically relevant rate—for instance, in alluvial areas because of deposition and river erosion, in arid regions and near sandy seacoasts because of wind erosion and deposition, and in humid regions through leaching.

A Different Classification

Clearly, then, the differentiation between exhaustible and inexhaustible resources indicates only a difference of degree. Even if changes of wants and of technology are excluded by assumption, the terms "exhaustible" and "inexhaustible" imply assumptions of the time pattern of revenues and costs as influenced by cumulative use and by physical variations of resources without use.

In our analysis it will be preferable to classify resources on a somewhat different basis. The two main classes will be called, respectively, "nonrenewable" or *"stock" resources* and "renewable" or *"flow" resources,* each of which in turn may be divided into subclasses.

The terms "exhaustible" and "stock" and the terms "inexhaustible" and "flow" are by no means identical. Many flow resources are exhaustible, and many stock resources are, economically speaking, inexhaustible.

Stock Resources

Resources are defined as "stock resources" if their total physical quantity does not increase significantly with time.[4] "Sig-

[4] Resources, as we know, are measured in appropriate physical units. These units must be chosen in such a way that variations in quality are taken into account. Variations in quality occur commonly in a given deposit at a given time—for example, in

nificantly" is defined from the economist's viewpoint. Strictly speaking, some stock resources may increase over time, but at a rate too slow to be economically relevant; for example, from the geologist's point of view, the formation of coal and oil may still continue.

Thus, with the total stock limited in quantity, each rate of use diminishes some future rate. Nevertheless, as has already been mentioned, use can be maintained indefinitely with some stock resources; this possibility depends on the quantitative relation between rates of use and stock. This relation depends in turn on revenue and cost conditions. However, the ends-means scheme that defines a resource usually implies limitations. For example, the economist is interested not in the stock of coal in the whole cosmos, but in the coal in a certain locality, at a certain depth, with a certain width of seam, and so on. Under these conditions, the quantitative limit on the stock implies also a limit on the time over which a stock resource can be utilized.

Figuratively speaking, stock resources are a reservoir without significant intake but with possible leakages. The level of the reservoir is lowered through each rate of leakage and through each rate of use. The level at which the rate of use becomes zero is an economic problem determined by the time pattern of revenues and costs, which change with cumulative use and cumulative leakage.

Classes of Stock Resources

For economic analysis, it is advisable to differentiate between two major classes of stock resources, namely, those which do not decrease significantly in time without use, and those which do. Many stock resources belong to the first class. Most metal ores *in situ*, coal, clays, and stones are examples. Oil and gas belong to the second class if changes take place through seepage and blowoff. Plant nutrients subject to leaching, refined metals subject to oxidation, and radioactive substances in the process of nuclear disintegration also belong to that group.

the thermal value of coal. Conceivably, such variations in quality may also occur over time.

Natural decreases of stock resources can technologically be influenced by human action. Here again, the problem arises as to whether it is economically feasible to do so. Leaching of plant nutrients, for instance, may be reduced economically through agricultural practices, whereas salinity control in arid regions under poor drainage conditions may not be economical. Metals may be economically protected against oxidation, whereas means to influence the disintegration of elements, although scientifically available, have not yet reached the economic stage.

Some stock resources may change from the first to the second class through use. This may happen, for example, to a potash mine after mining operations have started and are for some reason temporarily interrupted. The resource may deteriorate now because water that was originally kept out by protective strata of rocks enters through the mining shafts and dissolves the salts. This happened to potash mines in Alsace-Lorraine after operations were interrupted at the end of the First World War. At the end of the Second World War a high-grade but high-cost (as compared to South American ores) bauxite deposit in the United States was threatened with destruction when government orders were suddenly canceled after the resource was opened up; as with potash, the destructive agent is water, which dissolves clays and brings about cave-ins that render resumption of operations uneconomical. Similar problems exist in tin mining. Frequently, natural gas finds an avenue of escape through the drilling of an oil well.

Flow Resources

Resources are defined as "flow resources" if different units become available for use in different intervals. These successively available quantities constitute the "flow." The flow, without use, may increase or decrease continuously or discontinuously at either a constant or a varying rate. The present flow (which should not be confused with use) does not diminish future flow, and it is possible to maintain use indefinitely provided the flow continues.

Many flow resources can be stored. A given store of a flow resource can, of course, be treated as a stock. The availability of a flow for replenishing the stock, however, differentiates stored flow resources from stock resources proper. For example, on the basis of this differentiation ground water is usually a flow resource. Although the rate of natural replenishment may be small in relation to the total quantity stored in an aquifer, such a rate is usually significant in relation to the rate of use. Furthermore, the natural rate of replenishment may be augmented by artificial infiltration of surface flow. On the other hand, in some cases (connate water) ground water is a true stock resource. There is no natural replenishment, and surface flow may not be available for artificial infiltration.

Two economically relevant classes of flow resources may be differentiated according to whether the flow is or is not affected by human action under the present state of technology.

In the first class, human action in one interval does not significantly affect the flow in future intervals. Most flow resources in this class have a constant or cyclical flow, the duration of which for purposes of the economist may be assumed to be infinite. Solar and other cosmic radiation, winds, and tides are such resources. Figuratively speaking, in this class of flow resources only one cross section of the flow is under human influence.

In the second class of flow resources, on the other hand, human action in any given interval may decrease or may increase some or all future rates of flow. These changes of flow, in turn, affect the time pattern of revenues and costs and, therefore, future rates of use. Thus, more than one and possibly many cross sections of flow are under human influence. The rate of flow at cross sections "downstream" shows the effects of what has happened at various places "upstream." Examples will be given after a further classification, which is of great importance for the economics of conservation.

The Critical Zone

The second class of flow resources may be subdivided into two classes according to the existence or nonexistence of a *criti-*

cal zone in the decrease that may be caused, in rates of flow, by human action.

"Critical zone" means a more or less clearly defined range of rates below which a decrease in flow cannot be reversed *economically* under presently foreseeable conditions. Frequently, such irreversibility is not only economic but also technological. The decrease in the flow of animal and plant life, for instance, becomes technologically irreversible within a certain species if the rate of flow once reaches zero—that is, if the breeding stock is destroyed. Within a species, destruction of individuals that carry rare genes may also be important.

Even if the rate of flow has not reached zero, economic reversibility may disappear if highly complex ecological relations are affected—for example, plant associations and successions on some grasslands and forests. If overdraft of ground water has led to compaction of clay aquifers, restoration of storage capacity becomes economically impossible. (Ground-water storage capacity may, in the same way as surface storage capacity in reservoirs, be regarded as an important flow resource.) A decrease in soil productivity [5] can sometimes be reversed relatively cheaply if it results from depletion of plant nutrients; but if deep gullies have been formed which interfere with farm operations, or if all soil has been destroyed through erosion to bedrock, the economic reversibility of soil productivity, at least in certain uses—for example, cultivated crops—may have disappeared. Some scenic resources—for example, wilderness areas—may be spoiled irreversibly.

Economic irreversibility depends, obviously, on technology, wants, and social institutions. All of these are in constant change. In actuality, economic analysis deals with an open system. Present expectations of economic irreversibility, therefore, are subject to change and to great uncertainty. For example, it is by no means

[5] What is commonly called "soil" is a composite of many interrelated stock and flow resources. They are complementary in demand, as explained later. These resources can be used only through plants. As a practical first approximation, it is meaningful to measure the use of the composite resource "soil" in terms of the flow of plant production.

certain whether badly gullied farmlands, which under presently foreseeable conditions are not cultivable, may not be returned to cultivation under price relations and technology that may conceivably prevail at some future date. The implications of uncertainty in private and in social economics will be considered in detail later. Uncertainty of irreversibility, however, does not weaken the reason for the classification of resources proposed in this study: it is significant for economic analysis that for some resources a decrease of flow through human action carries the *risk* of irreversibility, whereas for others such a risk is not expected at a given time and place.

One aspect of the critical zone, as the concept is employed in this study, deserves emphasis: we are primarily concerned with the intertemporal relations between use rates of the same resource. The human actions, therefore, which irreversibly decrease the flow of a resource are primarily connected with the use of that same resource. Frequently, one resource is impaired irreversibly by using another resource. For example, in strip coal mining, soil usable for the production of crops may be impaired irreversibly in order to obtain the underlying coal; in hydraulic gold mining, agricultural and recreational resources of an alluvial valley may be impaired irreversibly in order to recover the gold. These are cases of resource interrelations, the various aspects of which will be referred to repeatedly.

Flow Resources without a Critical Zone

The flow of some important resources may be decreased (by human action) *without* a critical zone. An interesting example is precipitation. At a given interval over a given area, precipitation may be increased through cloud seeding (nucleation). The substances most used are dry ice and silver iodide.

Although the specialists are not in agreement on this point,[6] there is some evidence that increases of precipitation may be fol-

[6] Some specialists are of the opinion that a cloud loses less than 1 per cent of its moisture in the course of a heavy storm and that it is merged soon afterwards with air currents which restore what has been lost.

lowed by decreases in later intervals over other areas to which the clouds would have carried their moisture if no seeding operation had taken place. Furthermore, modification of precipitation might cause changes in temperature, wind direction, and wind velocities. These changes in turn might decrease precipitation in later intervals over the same area—besides other areas—over which precipitation was increased previously. There can be little doubt that modern technology will make it possible (if it has not already done so) to modify the flow of precipitation to a considerable extent. There is, however, no indication that such modifications will be irreversible.

The case of precipitation brings up an interesting possibility: it may become possible to modify also the flow of those flow resources which at present fall into the first class (cosmic radiation, tides, winds). Such modification, likewise, would scarcely be irreversible. Thus, if this possibility should actually come to pass, we would have only two classes of flow resources, both influenced by human action, but one without and one with a critical zone.

The services flowing from the utilization of producer and consumer durable goods may also be grouped in the class of flow resources without a critical zone. Although such goods may be classed among stock resources if attention is focused on them individually as use bearers, it is much more realistic to consider the productive services realized from them as flow resources. For technological reasons, productive services become available only over a number of intervals, and their flow can be increased, kept constant, or decreased. Considering durable goods of a certain type as a whole—as a species—a critical zone does not exist because they can be reproduced. In the utilization of an individual durable good a critical zone may exist if maintenance and repair is neglected.

The site values of special locations may also be grouped in this class. Such values may be impaired through use; but frequently they can be restored economically if use is changed—for example, as described in chapter 19, through zoning ordinances.

Table of Resource Classes

The classification of resources just given will prove of significance for the subsequent analysis, and may, therefore, be repeated in outline form:

I. Nonrenewable or stock resources.
   1. Stock not significantly affected by natural deterioration.
      Examples: metal ores *in situ;* coal; stones; clays.
   2. Stock significantly affected by natural deterioration.
      Examples: refined metals subject to oxidation; oil and gas in cases of seepage and blowoff; plant nutrients subject to leaching; radioactive substances (radium, uranium, plutonium, etc.) in process of nuclear disintegration.[7]

II. Renewable or flow resources.
   1. Flow not significantly affected by human action.
      Examples: solar and other cosmic radiation; tides; winds.
   2. Flow significantly affected by human action.
      a) Reversibility of a decrease in flow not characterized by a critical zone.
      Examples: precipitation; special locations that form

[7] Presently available information is not sufficient to classify with certainty atomic (nuclear) energy from the standpoint of economic analysis. There is some possibility that such energy, not unlike solar energy, may be classified as a flow resource. We refer to the possibility of operating continuous self-sustaining reactors. The potential economic importance of this possibility warrants a quotation from an expert in this field:

"One of the most exciting prospects in the development of nuclear reactors is the possibility that one may be able to consume fissionable material, obtaining energy from this process, and at the same time utilize the neutrons thus produced to generate a larger amount of fissionable material than was destroyed. This subject is now being actively pursued. Whether or not it will be possible to achieve is not yet clear. In order to be practical, it will also be necessary to develop highly efficient chemical treatment of the nuclear fuel which must be reprocessed from time to time. It is a very intriguing possibility, and since the amount of new material which can be produced will depend both upon the intrinsic rate of production in the reactor and the amount of fissionable material thus committed, our stock of fissionable material may prove to be an even more precious possession than it seems today."—Robert F. Bacher, "Research and the Development of Atomic Energy," *Science*, Vol. 109, January 1949, p. 7.

the basis of site value; services from a species of durable producer and consumer goods.

b) Reversibility of a decrease in flow characterized by a critical zone.

Examples: animals and plants; scenic resources.

Primary and Secondary Resources

Resources are frequently grouped into primary and secondary, the latter type being "derived" from one or several of the former. This classification depends on the viewpoint and the objective of the investigator—it is not a fixed characteristic of resources. In the physical as well as the economic sense most resources are dependent on others.

The flow of animal production and of wildlife may be called secondary resources with respect to crop production and natural vegetation; frequently, on the other hand, crop production and natural vegetation depend for their existence on domestic livestock and wildlife. Crop production and natural vegetation may be considered secondary resources with respect to a number of stock and flow resources usually grouped together as soil (see footnote 5); the formation and changes of soil, on the other hand, are greatly influenced by crop production and natural vegetation. Vegetation may be regarded as secondary with respect to precipitation; precipitation again may be regarded as secondary with respect to solar radiation; and solar radiation again secondary with respect to nuclear changes in the sun.

Stock resources and the first major subgroup of flow resources appear as primary resources for most purposes of analysis; the second major subgroup of flow resources comprises mostly secondary resources.

The terms "primary" and "secondary" will not be used in this study. Instead, the focus will be on the general problem of resource interrelations.

Relations between Resources

The relations between primary and secondary resources may be regarded as a special case of the more general problem of resource interrelation. This problem is of greatest importance for the economics of conservation. At this point only a few examples will be given. The economic consequences will be discussed in chapters 4 and 5 after the meaning of conservation has become clearer.

Relations between resources, in the same way as resources themselves, must be defined with reference to the ends-means scheme. When resources are used (in the sense of "produced," "supplied" [8]) and consumed (in the sense of "processed," "demanded"), *complementary* or *competitive* relations may exist. If such relations do not exist, resources are called "independent" —their relations are *neutral.*

Silver and copper ores are closely related through common production. Crops and livestock on diversified farms are similarly related. Under some conditions, use of land for forestry can well be combined with its use for recreation. Such relations may be called "complementarity in supply" (use, production).

A similar complementarity exists in demand (processing, consumption). For example, coal is related to iron ore because both are consumed together in the production of iron and steel. In agriculture, phosphorus, potassium, nitrogen, and many other plant nutrients, together with soil moisture, light, and other resources, complement each other in the production of plants.

On the other hand, resources may be related to each other because they compete. This competitiveness may also be either in supply or in demand.

Under some conditions, recreational use of wild lands may compete with forestry, ranching, and water production. Agricultural production competes with coal production on strip coal fields in midwestern and eastern states. Hydraulic gold mining competes with the production of fish and forage in some alluvial valleys of California.

A similar competitiveness exists in demand. Coal, oil, gas, and wood, or, in general, chemical energy, is related to water power and nuclear energy because they may compete with each other

[8] The terms "supply" and "demand" are generally employed in economics in connection with complementarity and competitiveness.

(can be substituted for each other) in many important lines of consumption. Animal power may compete with mechanical power in the demand by farmers; so may various types of carbohydrate feeds. In a multiple-use reservoir, storage space demanded for flood control at certain seasons may compete with storage space demanded for irrigation.

Some of these examples indicate that the type of the relations between resources—that is, complementarity and competitiveness—and the degree of these relations depend on the stages, the numerical values of supply and demand, that are considered. For example, the production of forage for deer (recreation) may be complementary to the production of trees (forestry) if the watershed is not too heavily browsed by a deer herd. However, if the watershed is overstocked with deer, competitive relations may come into being. Likewise, up to a certain size of herd, deer and domestic livestock do not seriously compete with each other on many ranges. However, after herds exceed a certain size, which depends on the capacity of a range for browsing and grazing, competitive relations may become serious.

The problem of changes in relations at different stages—that is, at different numerical values of demand and supply—requires a more precise definition of complementarity and competitiveness. For such definitions the second cross partial derivatives of supply (marginal cost) and demand (marginal revenue) functions are used. The reasons will be given when the relations between use rates of the same resource in different planning intervals are discussed in chapter 5 and the Appendix, section 3.

The relations between different resources have in principle the same significance and can be studied with the same theoretical tools as the relations between rates of use of the same resource in different intervals. This study is concerned with the latter, and care will have to be taken in our treatment that these two types of relations are kept apart and that the intertemporal ones are put clearly into the foreground.

Significant Implications

Some significant implications of the differences between resource classes, discussed on the preceding pages, may be noted briefly at this time in order to introduce general themes which will recur in a number of variations throughout this study.

A rather broad generalization may be suggested first: in the utilization of stock resources, the state of technology plays a dominant role; in the utilization of flow resources, economic and social institutions are especially important. What is the basis for such a general statement?

Stock resources are mainly used as raw materials for the production of durable goods and energy. Technological change is of great importance for these end products and for the productive processes leading to them. For most purposes of analysis, stock resources are primary resources and (except plant nutrients) their interrelations with other resources are nonbiological. This relatively (as compared with flow resources) simple state with respect to resource interrelations suggests that discovery and transformation of stock resources will be mastered more and more by advances in the natural sciences and will become more and more applicable economically by advances in industrial technology. Such an expectation is strengthened by recent applications of some advances of modern physics and chemistry (transformation of one element into another; availability of nuclear energy).

Flow resources of the first major subclass (class II, 1) cannot be depleted. Depletion of other flow resources does not create serious economic and social problems if a critical zone does not exist (class II, 2, *a*).[9]

Flow resources with a critical zone (class II, 2, *b*) are the only group of flow resources which gives rise to serious social and economic problems of depletion. These resources serve mainly

[9] The economic and social problems created by the utilization of these resources are more connected with the distribution of revenues and costs among individuals than with the depletion of the resource for a social group as a whole. An example is the serious problems which will be created by modification of precipitation.

for food, clothing, recreation, and aesthetic enjoyment. Except in clothing, technological change has done comparatively little to alter these wants or the means of their satisfaction. Flow resources are secondary resources, and (within class II, 2, *b*) are related to each other (and to plant nutrients) through highly complex biological processes. Although technology affects these processes, such influences cannot be expected to be great enough or fast enough to lessen the economic and social significance of their depletion. On the other hand, use of these resources is rather sensitive to economic and social institutions—for example, those affecting land utilization.

The differences between resource classes just indicated have several significant implications.

First, in stock resources, uncertainty concerning technological change is a much greater impediment to planning by individuals or governments than in flow resources.

Second, the confidence that technological progress will make social action in the interest of the distant future unnecessary appears justified for some important stock resources—for example, those used in the production of energy.

Third, such confidence is only too often unwarranted complacency for flow resources with a critical zone (class II, 2, *b*); there is great danger that such confidence may be carried over from the field of stock resources.

Fourth, public conservation policy—that is, a change in the institutional framework—is especially promising as well as urgent for flow resources.

Finally, recent important technological advances in the production of synthetics and in the transformation of energy (for example, the possibility of obtaining nuclear in addition to chemical energy, progress in the field of synthetic fuels, new motors based on jet propulsion) do not obviate an inquiry into the economics of conservation. On the contrary, these advances underline the necessity for clear thinking about the meaning of conservation. This meaning will be explained in the next chapter.

CHAPTER 4

MEANING OF CONSERVATION

Conservation Does Not Mean Nonuse

On the basis of the preceding classification of resources, we can now approach the meaning of conservation somewhat more closely.

The conservation of stock resources in the sense of keeping the stock permanently undiminished and unimpaired is clearly meaningless. If conservation were understood in this sense, rates of use would have to remain zero, and the resource character of that portion of the environment which is considered would be lost.

Such a meaning of conservation would, furthermore, be highly misleading for flow resources. Conservation of a flow is economically meaningful only if its use is considered. Conservation in itself cannot mean nonuse. Misunderstanding on this point has had unfortunate consequences. For example, in the conservation clause of the constitution of the state of New York (Art. VII, Sec. 7, added in 1894) relating to the state forest preserves, use was prohibited.[1] Difficulties were created for effective forest administration and for the general public acceptance of conservation.

[1] This clause reads in part: "The land constituting the Forest Preserve shall not be leased, sold, or exchanged, . . . nor shall the timber thereon be sold, removed, or

Is Conservation "Wise Use"?

By many people, to be sure, use is considered when the term "conservation" is employed. By some, conservation is interpreted as "wise use"; by others as "the greatest use to the greatest number over the greatest length of time." Such interpretations of conservation have little meaning and are of no value for economic analysis.

Nothing is gained by interpreting conservation as "wise use" if the adjective "wise" is not defined.

"The greatest use to the greatest number over the greatest length of time" refers to three objectives which are obviously in conflict with each other. Without indicating the way in which these conflicts could be reconciled, such a "definition" of conservation merely adds to the existing confusion.

There is another interpretation of "conservation" which is based on a static state of use. This meaning deserves a more detailed consideration.

Is Conservation a Static State of Use?

By many authors, including economists, "conservation" is defined as keeping use at a constant rate—that is, as "sustained yield," to employ a term which has become popular, especially in forestry.

Such a definition of conservation is meaningless for stock resources. As a rule, use of such resources cannot be sustained indefinitely (chapter 3).

For flow resources, such a definition would exclude all those cases in which an attempt is made merely to slow down rather than to stop a decrease of flow and use. Alternatives between different rates of decrease in use may be relevant economically when maintaining use at a constant rate is not. A farmer may be content with slowing down a decrease of yields from cultivated crops.

destroyed." After the storm of November 25, 1950, heavy blowdowns created such a fire hazard that a special bill was enacted by the New York legislature, which provided for the "removal, use, or sale of fallen or seriously damaged trees." This bill still prevented the cutting of standing timber which blocked access to fallen patches!

Such decrease frequently occurs for those types of soil (prairie and pampas soils, peat soils) the productivity of which is the result of long periods of undisturbed accumulation of organic matter. It is technologically difficult, and seldom economical, to keep such soils as productive as during the first few years after breaking the virgin sod. But it may be economical to slow down the decrease in crop yields; a farmer who does that would say that he was practicing conservation. In general, as discussed in chapter 18, conservation is economical only when slowing down rather than stopping natural (geologic) erosion and leaching is attempted.

Furthermore, for flow resources, different use rates, each constant over time—that is, different levels of sustained yield—are frequently economic alternatives. Whether any rate of use should be maintained indefinitely may be less relevant economically than what such a rate should be.

Under certain physical conditions in agriculture, forestry, and grazing, depletion of resources reaches a certain base level. After a very low level of yields is reached, it may be sustained indefinitely by the natural accretion of plant nutrients without special costs for maintenance. But there would be little meaning in calling such a static state "conservation." On the other hand, an increase in the level of sustained yield may well be called "conservation," although an alternative lower level would also be maintained indefinitely.

As will be shown presently, not *all* slowing down of a decrease in use rates and by no means *all* increases of use rates are conservation. Proper use of the term depends on the time distribution of these changes. This may well be emphasized because the concept "conservation" has recently been used by government agencies and others in the all-inclusive sense of increasing use. No special term would be needed if conservation meant merely an increase of use.

As was emphasized earlier, rates of use in economic reality are rarely constant over time even if changes of wants and of technology are excluded by assumption. The reasons are, first, the effects of one rate upon future rates through revenues and costs and, second, qualitative and quantitative variations in resources that may take place without use. Constancy of rates of use is merely a hypothetical special case in resource utilization. For economic reality, such a special case has little significance. This is true both for a study of historical facts and for a consideration of private and social objectives. True, with such a restriction of the concept, the tools of simultaneous economic analysis can be employed—but only at the expense of oversimplifying or evading some essential theoretical problems, and of impeding practical use of the theory in economic reality.

A Dynamic Interpretation of Conservation

As indicated in chapter 1, the concept "conservation" is tied in this study to a particular aspect of use: its intertemporal distribution. Conservation is concerned with the *when* of use.

More specifically, "conservation" and its logical corollary but economic opposite, "depletion," are defined in terms of *changes* in the intertemporal distribution of use. In conservation, the redistribution of use is in the direction of the future; in depletion, in the direction of the present.

Conservation always implies comparison of two or more time distributions of use. We may compare expected use if new practices are adopted with what use would have been if the old practices had been continued. We may compare several utilization plans merely by calculation, by budgeting. Or we may compare actual time distributions of use during different periods of history or in different geographic locations.

The terms "in the direction of the future" and "in the direction of the present" could be defined simply on the basis of the time sequence of increases and decreases of use rates. This would be satisfactory if mere qualitative differentiation between conservation and depletion were sufficient and if increases and decreases occurred without alternations over time between the two or, as a special case, if all changes over the whole planning period were of the same sign. These conditions are fulfilled in many practical problems of conservation. In most parts of this study, therefore, a simple directional (with respect to time—

toward the future or toward the present) interpretation of conservation and depletion is sufficient; and the reader not particularly interested in economic theory may pass over the following section.[2]

Quantitative Measurement of Conservation

A theoretically satisfactory definition of conservation and depletion must be applicable to all conceivable alternations over time between positive and negative changes in use rates over the whole planning period,[3] and must provide (for example, in problems of conservation policy) for comparing different degrees of conservation and depletion—that is, for quantitative measurement.

This purpose can be achieved by considering the aggregate of all changes, having due regard to their sign and using as weights their distances from that interval which is regarded as the starting point in planning or studying a redistribution of use rates. The starting point may be called "the present," in order to avoid cumbersome language. In this study the first power of distance rather than a higher power or a more complicated function of distance is used for weighting.

The choice of weighting used in this study may be criticized as arbitrary. However, the particular function of time used for weighting has no effect upon the subsequent argument. Theoretically the function used here is not inferior to any possible alternative. It has the practical merit of being the simplest and of being identical with the function of time that is generally involved in economic analysis, for instance, in compounding and discounting. Essentially the issue is an index-number problem

---

[2] As indicated in chapter 2, one objective of this study is to reach both the general reader and the professional economist. To this end, a two-level treatment will be used repeatedly.

[3] How important it is to compare changes of use rates over the whole planning period (not merely over sections of this period) may be illustrated by an example: in mature stands of forests, an increase in utilization near the present is frequently a means of conservation. Growth is greatly stimulated and, although use may be lower for some intervals after the cutting (as compared with the alternative of deferred cutting), use will be greater later on.

with the similar difficulty of finding the "best" method of weighting.

We may then quantitatively define "conservation" as changes in the time distribution of use rates of individual resources in which the aggregate weighted change in use rates is greater than zero. Correspondingly, we have depletion if the aggregate weighted change is less than zero. If the aggregate weighted change is equal to zero, there is neither conservation nor depletion, and the actions considered may be called "neutral." Further, we may define and measure quantitatively the degree of conservation or depletion by the ratio of the aggregate weighted change in use rates to the aggregate weighted use rates existing without the redistribution in question. The degree of conservation may be most conveniently expressed in per cent.[4]

State of Conservation

The terms "conservation" and "depletion" as just defined refer to a change of use rates, not to a given (unweighted) time distribution of use rates.

Any time distribution of use rates may be regarded as the *result* of conservation or depletion if compared with a different distribution. In the interest of shortness, we will call a given time distribution of use rates a "state of conservation." [5] ("State of conservation" may be regarded as synonymous with "state of depletion"; the latter term, however, will not be employed in this study.)

An analytically important special case of such a state of conservation will be discussed below as the "optimum state of conservation."

As a consequence of this terminology, it may be noted that a change in the state of conservation (including a change in the optimum state of conservation) may mean either conservation or depletion.

[4] For sake of precision, a mathematical formulation of these definitions is presented in the Appendix, section 1. The reader may also refer to S. V. Ciriacy-Wantrup, "Taxation and the Conservation of Resources," *Quarterly Journal of Economics*, Vol. 58, No. 2, February 1944, pp. 157–195.

[5] In the sense of "resulting from" or "with respect to" conservation.

"State of conservation" is merely a convenient shorthand expression for "a given state in the intertemporal distribution of use rates."

Conservation Ex Post *and* Ex Ante

The term "conservation" may be applied to both actual and hypothetical changes in the intertemporal distribution of use. Similarly, the term "state of conservation" may be applied to both observed and planned time distributions of use rates.

There are two main reasons why conservation in the statistical or historical sense—conservation *ex post*—must be differentiated analytically from intended—*ex ante*—conservation.

First, although the intentions of individuals and of governments are obviously of greatest importance for the results obtained, it frequently happens that conservation (or depletion) occurs as an unintentional by-product of private or public planning in other spheres; likewise, sometimes depletion may be the result of attempts to bring about conservation, or vice versa; why and under what conditions such results occur is of great interest in a study of public conservation policies.

Second, changes of social institutions, of technology, of wants, of income distribution, and of many other economic forces may usefully be analyzed from the standpoint of whether they result in conservation or depletion regardless of any intentions of planning agents. This is the same reason which led us in chapter 3 to a differentiation between resources *ex ante* and resources *ex post*.

Measuring Not Affected by the Money Veil

Use, as we know, is always expressed and measured in appropriate physical units—tons, calories, kilowatt-hours, acre-feet. Such measurement implies that the quality of the resource is assumed the same—in other words, that variations in physical characteristics are appropriately taken into account through the unit of measurement. For example, in measuring coal resources the appropriate unit may be a thermal one rather than a ton, if variations of quality occur.

Thus, changes in the monetary value of use rates do not affect the measurement of conservation and depletion. This is worth noting, because for many people the money veil tends to obscure changes in the state of conservation. Increases in the unit value (price) of a resource due to its use or to more general causes of price changes, and the resulting comparative stability or even increase of total income and capital value of the resource, tends to blind both resource users themselves and their governments to the true state of affairs, and to delay remedial action.

Measuring Conservation of a Composite of Resources

Our definitions relate to individual resources. This is important to keep in mind if relations between different resources are close and complex (chapter 3). In such cases it is easy to regard use rates of a whole composite of resources as an indication of the state of conservation of one of its components. For some purposes of analysis little harm comes from such convenient first approximations, provided the investigator knows what he is doing. However, very misleading conclusions may be drawn if the investigator forgets that he deals with a composite of resources, and that an increase in rates of use from the composite may be perfectly compatible with a decrease in the contribution by one or even several of the components. An example may illustrate this point.

Soil, as we know, is a highly complex composite of resources the individual components of which are (among others) plant nutrients, water, and many species of higher and lower plants and animals. In addition, soil productivity (the flow of crop yields) involves such resources as solar radiation, one or several species of crop plants, human effort, and equipment. By many investigators crop yields are regarded as an indication of the state of conservation of the resource "soil." Stability or even increases of crop yields—because of an increased genetic potential of crop plants, or of a cycle of wet years (in nonhumid areas), or of changes in cropping practices—may obscure the effects of sheet erosion, of leaching of plant nutrients, of disappearance of organic matter in the A-horizon, and of other forms of soil depletion. Loose thinking of this type may, therefore—as in the case of the money veil referred to above—lead to complacency of resource users and governments, and prevent remedial action until it may be too late.

Accomplishing Conservation

Conservation and depletion as defined above refer to changes in the time distribution of physical outputs. To accomplish this, it is usually necessary to change the time distribution of physical inputs.

To defer output (with no change of input) is only one way of accomplishing conservation. Although this kind of "waiting" is commonly stressed in the literature on conservation, it scarcely ever occurs in its pure form in reality. Conservation is, however, frequently accomplished by changing inputs without any deferment of output (as compared with the output of an alternative plan).

Technologically, different ways of accomplishing conservation are closely intertwined. For economic analysis, these different ways represent different combinations of dated productive services to obtain a desired time pattern of use rates.

Conservation and "Best Knowledge"

When the concept "resource" was discussed, reference was made to the level of technology—the "state of the arts" of economic texts. Obviously, individual resource users do not plan the time distribution of inputs and outputs on the basis of a uniform "level" or "state" of knowledge. On the contrary, in a given social group, individual differences in knowledge are usually great and important for production plans.

Differences in knowledge within a given social group, barriers and lags in tendencies toward equalization of these differences, and possibilities of overcoming such barriers and lags are of greatest importance in the social economics of conservation (Parts IV and V). In the economic analysis of the behavior of individual resource users, however, it is convenient to assume that they plan production on the basis of a given "best knowledge." Accordingly, this assumption will be made in the private economics of conservation (Parts II and III).

"Best knowledge" may be defined for the purposes of this study as the knowledge available in a given social group at a given time. Best knowledge, therefore, is not influenced by the better knowledge available to the observer or to different social groups or to the same social group at a different time. Neither does "best knowledge" imply that uncertainty concerning the future is removed or even diminished.

This definition of "best knowledge" allows us to isolate and to study the problems of education as an important tool of conservation policy.[6]

This definition has also some bearing on the term "waste," which is so frequently employed in the literature on conservation.

Conservation and "Waste"

The concepts "conservation," "depletion," and "state of conservation" carry no connotation of efficiency or waste. Any change in the time distribution of use rates and any state of conservation may be called "wasteful" if not based on best knowledge. However, nothing is gained by such a general use of the term "waste." Furthermore, as just stated, in the private economics of conservation our analysis will proceed under the assumption that individual planning agents employ given best knowledge.

In the literature on conservation the term "waste" is usually associated with the depletion of flow resources; it means that such depletion is regarded as either privately or socially uneconomical. In this sense, however, conservation likewise may be wasteful. A few illustrations must suffice at this point.

Conservation that would try to hold agricultural productivity of a virgin soil with distant markets at the original level or try to restore it to that level may be wasteful in terms of human effort, equipment, and natural resources (lime, fertilizer, fuel). A forester who would hasten the recovery of a cutover forest in a

[6] However, conservation is not a necessary result of better education, and education is not a cure-all for social problems of resource depletion (chap. 19).

young country through planting instead of waiting for natural revegetation may be wasteful because his efforts may better be directed toward fire protection, improved systems of logging, and the like, over a larger forested area. A mine owner who installs expensive permanent equipment in order to increase his recovery rate or to extend the life of his mine may be wasting labor and capital which he could more advantageously employ in alternative enterprises. Under some conditions, depletion of local ground water may enable a community to grow strong enough to import distant surface water, whereas conserving ground water would necessitate limiting its development.

Apparently, there is a most economical way of accomplishing conservation and an economic limit in conservation. Somewhere, in conservation, an economically optimum distribution of rates of use over time is reached. This distribution we will call the "optimum state of conservation." Its meaning is highly complex, and will be explored more fully in the next chapters.

Conservation and Investment

It may be well to note the differences between "conservation" and "investment," and between "depletion" and "disinvestment." So much of current economic analysis stresses investment policy that an understanding on this point appears essential.

"Conservation" and "depletion," as we know, refer to physical changes in the time distribution of use rates of individual resources. "Investment" and "disinvestment," on the other hand, have no necessary relation to use rates of individual resources. Although these two terms are sometimes used ambiguously, most economists will agree that "investment" and "disinvestment" refer to value changes in total capital of persons, firms, or whole social groups as a result of differences between income and consumption—and not to physical changes in the use of individual resources.

Often, investment results in depletion, and disinvestment in conservation.[7] For example, investment connected with the tech-

[7] This is true also if the term "real" is employed in connection with investment and disinvestment. This term is used if values are corrected for price changes over time.

nological changes mentioned in chapter 1 has frequently led to depletion. Conversely, more conservative stocking of ranges, replacement of cultivated row crops by a permanent grass and livestock economy, proper spacing of oil wells, restrictions on fishing and hunting, and many other private and public conservation practices may necessitate disinvestment in private enterprises and in whole industries.

Even if the terms "investment" and "disinvestment" are employed, not in the scientific sense just indicated, but in the popular sense as referring to individual resources, they are by no means identical with the terms "conservation" and "depletion." For example, investment in oil wells, timberlands, or mines through purchase is in itself neither conservation nor depletion. What matters is the change, if any, that is brought about in the time pattern of use rates by the change in ownership.

Decreases in the unit values (prices) of an individual resource may lead to disinvestment—that is, to a decrease in capital value—and, at the same time, to a change in the utilization plan in the direction of conservation. An increase of prices may lead to an increase in capital value and nevertheless to depletion. Whether conservation or depletion occurs depends, as will be further explained in chapter 9, on how the changes of prices (price expectations) are distributed over time and on how rates of use in different planning intervals are related through revenues and costs.

A differentiation between "investment" and "conservation" and between "disinvestment" and "depletion" is important not only formally for economic theory, but also operationally for entrepreneurial decisions and public policy. Discussion of resource problems in terms of investment and disinvestment—that is, with reference to capital values and aggregates of different resources—tends to deflect scientific and popular attention from the technological, economic, and social problems of individual resources that are related to their physical characteristics, and also from those problems, abundant in conservation policy, that can only partly be analyzed in terms of market values.

A Maximum of Cumulative or Sustained Use

Our definitions imply that "conservation" and "depletion" have no necessary relation to changes in the *unweighted* aggregate of use rates in all intervals—that is, to changes in cumulative use. This is in conformity with popular terminology. Cumulative use of deteriorating stock resources (class I, 2) is often greatest if the stock is used up in the shortest possible time. Examples are natural gas and oil where losses of gas cannot be avoided, and already produced metals that unavoidably deteriorate to some extent through oxidation. The term "conservation" could hardly be applied to changes in the utilization plan that would aim at highest cumulative use in such cases. On the other hand, economizing in the use of strategic metals and stockpiling them for emergency would generally be called "conservation" even though the stored metals may deteriorate.

But this situation is somewhat different for those flow resources which are affected by human action (class II, 2). Here all those actions which are directed toward sustaining otherwise decreasing rates of use—toward sustained-yield management—or toward increasing a sustained use are conservation according to our definitions. Such actions are conservation until a maximum sustained use is reached. Maximum sustained use and maximum cumulative use are approximately identical if a sufficient time is considered. Here again, our definitions are in conformity with popular terminology.

Maximum Sustained Use as an Economic Objective?

Maximum sustained use is sometimes claimed as a general objective in private conservation and in public conservation policy—for example, in forestry and fisheries. The fisheries of the high seas may serve as an illustration because here the economic problems are simpler than in forestry.

In the high-seas fisheries, maximum sustained use is a meaningful economic objective only under the following three assumptions: First, there is a maximum harvest (use) which can be taken periodically without affecting the natural flow. Second, it is not economical to increase or stabilize (in cases of natural decreases or cyclical variations) the natural flow and the corresponding harvest by inputs—for example, by improvement of habitat, artificial propagation, feeding, and predator control. Third, costs for harvesting and demand for products are such that it is not economical to take less than the maximum harvest under natural flow conditions.

It is conceivable, although not likely, that these assumptions may be fulfilled in reality. But it would be a mistake to conclude that the economic objective in conservation could generally be defined in terms of a physical-biological maximum. Instead, we are dealing with a socioeconomic maximum the meaning of which will be developed in chapters 5 and 6.

PART II

THE PRIVATE ECONOMICS OF CONSERVATION

THE OPTIMUM STATE OF CONSERVATION

CHAPTER 5

CONSERVATION DECISIONS

Private Economics Considered First

The economic and social issues in conservation differ between private individuals and society. It is advisable, therefore, to treat conservation in private and in social economics separately. We may focus on the former first because most private issues are simpler, and because in a private-enterprise economy like that of the United States, both the need and the possibilities for public action are based on an understanding of the behavior of individual resource users and of the way their behavior is influenced by their environment—that is, the way it is influenced by "economic forces."

Such an understanding is required not only for policy makers and their advisors. In a field in which continuity is of utmost importance, a continuous public demand for legislation and enforcement is a requirement for action by a democracy. A continuous and articulate demand for conservation policy requires that a considerable portion of the electorate become aware of the economic issues involved. Likewise, implementation of conservation policy is aided if those immediately concerned, the individual resource users and their political representatives, understand the relations between their own behavior and the interests of the social group as a whole.

In this part of our study (Part II) the behavior of private resource users with respect to conservation will be considered. The following part (Part III) will deal with the economic forces that affect conservation in private economics. An analysis of these forces is the core of our study. On the basis of this analysis, we will then turn to the social economics of conservation (Parts IV and V).

Conservation Decisions and Production Planning

As we saw in chapter 4, conservation and depletion *ex ante* are expressed through changes in the series of use rates planned over time or, for short, through changes of the production plan. Most farmers, foresters, mine operators, and oil producers plan use for more than one interval. This must be done because the rate of use in one interval is related through revenues or costs, or both, to rates in other intervals. To plan only one rate would be uneconomical.

The necessity for an integrated production plan extending over time, and, therefore, the economic significance of conservation decisions will now be considered in more detail. First, let us define more clearly the types of economic relations between use rates and show why these relations require an integrated production plan. We will then illustrate these relations through examples from agriculture, forestry, mining, and the oil industry.

Relations between Use Rates Defined

In chapter 3, the relations between use rates of the same resource in different planning intervals were considered in order to analyze the effects of cumulative use upon revenues and costs. Somewhat later in the same chapter, the relations between different resources through supply and demand were discussed. The terms "complementary" and "competitive" were introduced with respect to these latter relations, and it was indicated that the same theoretical tools can be used for the analysis of both kinds of relations. Let us now define the terms "complementary" and "competitive" more precisely for the intertemporal kind of relations which, as pointed out before, are the main interest of this study.

When we say that two rates of use of the same resource in different planning intervals are related through revenues or costs, we mean that changes in one rate will affect the revenues or costs of the other.

Clearly, in order to measure such effects, the other rate must be kept constant; otherwise one would not know to which change of rates the change of revenues or costs is related.

Another requirement was implied when changes of resource relations at various stages were mentioned: at different numerical values—in physical units—of the two rates, changes of revenues or costs may be quite different. For example, up to a certain rate of use (in terms of tons of grass harvested or animal units pastured), grazing of pasturelands in one year may not increase the costs of taking the same harvest the next year. However, from a certain stage onward, an increase in the rate of use will require costs for irrigation, fertilizer, rotation, or other practices of pasture management, if the harvest the next year is to remain unchanged. If the effects of changes in one rate upon revenues or costs of another are to be measured for certain numerical values of these rates, we have to measure the changes that take place in the increase or decrease of total revenues or costs brought about by the last unit of the rates considered. In the terminology of the economist, we have to measure changes in *marginal* revenues or costs.

We may then define relations between two rates of use in different planning intervals as complementary in revenues (or net revenues) if an increase in one rate increases the marginal revenues (or net revenues) of the other rate. Relations are defined as competitive in revenues (or net revenues) if an increase in one rate decreases the marginal revenues (or net revenues) of the other. Relations are defined as complementary in costs if an increase in one rate decreases the marginal costs of the other. Competitiveness in costs prevails if an increase of marginal costs takes place. The relations are called "neutral" (or rates independent) if no changes result. A parallel set of definitions is used for the relations of different resources through demand or supply.[1]

Factors tending toward both complementarity and competitiveness, in costs as well as revenues, may be present in the relations between use rates in two different intervals. It is the net effect of all such factors which influences the planning agent when he tries to realize his economic objective in production planning, as discussed in chapter 6. For economic analysis, however, the effects of various factors may be separated—as in this chapter.

Theoretically, relations in revenues or costs between use in different intervals of the same production plan may vary—that is, some use rates may be competitive with each other; others may be complementary or neutral. In actuality, the situation is usually simpler. In many cases, all rates of use are independent in revenues. In some types of costs complementary relations prevail over the whole production plan. In other types of costs all rates are usually competitive. It is well, therefore, to study these relations somewhat further with a view to understanding their practical significance. This will be done first for relations through revenues and then for those through costs.

Relations through Revenues Depend on Market Form

All production implies that there are future wants to be satisfied, or, from the producer's standpoint, that future revenues are expected. However, revenue expectations, by themselves, do not require an integrated production plan. They are merely necessary conditions. An integrated production plan is required only if rates in different planning intervals are related through revenues or costs. As examples of interrelations through revenues we mentioned storable products and scrap in chapter 3. The practical significance of these examples is dependent on and restricted by

---

[1] In mathematic terms, these definitions are expressed by the second cross partial derivatives of the total revenue, total net-revenue, and total cost functions. They are thus expressed in the Appendix, section 3. The functions used there are in terms of present values, that is, after discounting. However, for the discussion of the text above, it is immaterial whether the functions are in terms of present value or future value. The difference between these values will be discussed later (chap. 7).

the form of the market in which resource users dispose of their products. Accordingly, the market form becomes significant for an analysis of conservation decisions.

Economists differentiate between a number of market forms. Two types of market forms interest us at this point. We will call them "pure competition" and "monopolistic conditions." [2] "Pure competition" is defined by stipulating that individual resource users do not expect that the quantities they produce—that is, rates of use in various planning intervals—will influence the prices of their products (or that the quantities of productive services they demand will affect the prices of such services). Under monopolistic conditions they expect such an influence. Under pure competition, product prices expected in different planning intervals are identical with marginal revenues expected for the same intervals. To regard prices as independently given is, therefore, a different way of saying that differently dated use rates are independent in revenues.

In actuality, pure competition is approximated for most users of flow resources—that is, for agriculture, forestry, grazing, and fisheries. In these industries, therefore, we are concerned largely with relations through costs and can disregard (neutral) relations through revenues, at least so far as we are interested in individual planning (conservation *ex ante*).

On the other hand, in the utilization of stock resources—metal ores, oil, fertilizer materials—monopolistic conditions are fairly common. Here, the relations through revenues may require an integrated production plan even if relations through costs could be neglected. Monopolistic conditions, therefore, are of considerable significance for resource conservation. It is advisable, however, to postpone a more detailed consideration of this problem until the economic behavior of individual resource users and the effects of economic forces upon their conservation decisions have become clearer. We will consider the relations between conservation and monopolistic conditions in chapter 14. At this point, we

[2] This term is intended to include market forms referred to in economic literature as "monopolistic competition," "oligopoly," "dyopoly," "monopoly," and the parallel phenomena on the buyer's side ("monopsonistic" conditions).

can focus on the relations of use rates through costs. Competitiveness in costs may be taken up first.

Competitiveness in Costs

Several examples for the relations between use rates through costs were mentioned in connection with the effects of cumulative use. All these relations were competitive—that is, an increase of use in one planning interval increased marginal costs in other intervals.

Competitiveness in costs between use in intervals nearer the present and use in intervals at greater distance is generally recognized in the literature on conservation. Such competitiveness is the major reason why conservation can be accomplished through decrease of present use rates in favor of other rates further removed in the future—that is, through "waiting" for output. Emphasis on competitiveness in costs is justified because this relation is frequently not sufficiently taken into account by individual resource users in production planning.

In farm accounting, for example, an increase in future costs caused by soil depletion, overgrazing, and overpumping is generally taken too lightly. This is partly due to the money veil and the composite character of the resource soil as explained earlier. It can partly be blamed on the laws and customs existing in the fields of land taxation, credit, and tenure. These social institutions treat the land as if it were a fixed "indestructible" resource. Such treatment is fostered by many texts in agricultural economics and farm management.

On the other hand, in forestry, mining, and oil drilling, allowances in taxation, credit, and tenure laws for the effects of present use upon future use (treatment of "depletion," "maintenance," and "development") have at least created an awareness of the problem—although the laws themselves, as we shall see in chapters 10 to 13, generally fall short of giving adequate consideration to the importance of competitiveness in costs.

Complementarity in Costs: Sunk Costs

Competitiveness in costs is not the only reason for an integrated production plan. Usually, complementarity in costs is even more important.

Complementarity in costs between use in intervals nearer the present with use in intervals at greater distance is largely a result of what we may call "sunk" costs.

Sunk costs include, first, costs which in short-run instantaneous (simultaneous) economic analysis are known as "fixed" and "lumpy."[3] Fixed and lumpy costs are sunk over time because the technological characteristics (usually mentioned in economic texts as indivisibility, immobility, and specialization) of necessary "factors" which give rise to these costs would prevent economic use of their potential productive capacities if such use were confined to any one planning interval. Such factors are, for example, buildings, transportation facilities, machinery, equipment, trees, and breeding stock.

Sunk costs include, second, costs which in instantaneous economic analysis are termed "variable." In agriculture, for example, the costs of fertilizer, feed, and labor are sunk over periods of gestation. All these inputs are potentially useful in obtaining revenues over more than one planning interval. Similar conditions prevail in other resource industries. For example, considerable costs for labor and materials must be expended to open up a mine or to drill a well before production can start. Such expenditures are potentially useful over many planning intervals. In order to reap the potential revenues from the expenditure of sunk costs, production has to be planned over time. Additional future costs necessary to reap these revenues may be called "recovery" costs.

Sunk costs play a role in maintaining production over time

[3] This means "fixed" and "lumpy" with respect to instantaneous variations of use. Lumpy costs differ from fixed costs in the degree of fixity. Fixed costs proper do not vary at all with instantaneous changes of use over the range considered. Lumpy costs are fixed over a part of the range that is large in relation to the whole range considered; they vary discontinuously. Frequently, lumpy costs are irreversibly variable: they are variable with increasing use, but resist variation with decreasing use. In such cases, if use is reduced, lumpy costs behave like fixed costs.

similar to that of fixed and lumpy costs in maintaining production over a certain range in instantaneous economics. Likewise, recovery costs have economic characteristics similar to those of variable costs of instantaneous economics.[4]

Complementarity in Costs: Opportunity Costs

Another cause for complementarity in costs is that it is often difficult for the planning agent and his family to shift to other types of employment. In terms of opportunity costs this difficulty is the same as the existence of sunk costs just discussed; it has the same consequences for maintaining production over time. However, the opportunity costs of the planning agent and his family are better treated separately. First, indivisibility, immobility, and specialization of the planning agent and his family (and, frequently, his workers) differ in their origin from similar conditions affecting other productive services. In the latter case these conditions are largely technological, whereas in the former they are largely institutional. Second, opportunity costs of the planning agent himself are analytically different and more complex than the costs of productive services he employs.

How these differences in relations, both in costs and revenues, affect the optimum shape (in terms of rates in various intervals) and extent (in terms of number of intervals) of the production plan will be discussed in the following chapter. Before this is done, it may be well to illustrate relations in costs by examples from agriculture, forestry, mining, and oil production.

Illustrations from Agriculture

In agriculture, sunk costs are particularly important. Investment in fixed factors such as improvements (buildings, drainage and irrigation systems, transportation facilities, trees), equipment, and breeding stock loom large. Variable inputs are sunk over various periods of gestation: one year in the production of

---

[4] S. V. Ciriacy-Wantrup, "The Relation of War Economics to Agriculture, with Particular Reference to the Effects of Income and Price Inflation and Deflation," *American Economic Review*, Vol. 30, No. 1, Pt. 2, Supplement, March 1940, pp. 366–382.

field crops; from one to four years in livestock production, if breeding stock is not increased; from one to eight years, if breeding stock must be increased in order to increase market supply; from five to ten years in the production of bearing trees.

The farmer himself is not freely mobile because knowledge of particular local conditions is an important factor of success in agriculture, and because family and other social connections in a particular area are more potent ties in farming than in most city occupations.

Relations between use rates through competitiveness in costs are numerous and significant. The degree of stocking in one year affects decisively the economics of grazing the next year. The efficiency of livestock in transforming feed input into milk, meat, and wool depends on the care taken in breeding and feeding during youth. Lowering of the water table through excessive pumping increases future irrigation costs. Soil depletion in its various forms (removal of plant nutrients through crops, leaching, sheet erosion, gullying) or, conversely, soil conservation through various practices (terracing, strip cropping, contour farming, better rotation, permanent grassland, or forest) affects future costs decisively. As special cases of these effects, changes in the costs of credit (because of changes in the value of assets which can serve as security) and in the scenic and aesthetic attractions of farm life may be mentioned.

Illustrations from Forestry

A forest owner faces the same necessity for production planning over time as does the farmer. If the forest owner harvests and processes his timber himself, he has high sunk costs in lumbering and sawmill equipment, buildings, and transportation facilities. Gestation periods for variable inputs are particularly long.

Although it is probably easier for a forest owner than for a farmer to find employment elsewhere, his capacities are not perfectly divisible, mobile, and unspecialized.

The effects of use rates in one interval upon increasing costs in future intervals are highly important. The forest owner must consider how many seed trees or seed strips should be left; how far he should protect existing young growth while harvesting the mature timber; how far he should reduce future fire hazards through disposal of slash; when and how many improvement cuts he should take; in what intervals and within what diameter limits he should harvest the mature trees.

Illustrations from Mining

A mine operator has much sunk investment in shafts, dredges, transportation facilities, air-conditioning systems, structures to support the overhang, ore-improvement plants, and workers' housing. Sunk costs for opening up a mine are frequently high.

Personal immobility is less important for him as planning agent than for a farmer or forest owner (though it may be important for his workers).

On the other hand, relations between use rates through competitiveness in costs are highly important. The decision of how much of a coal seam should be left in pillars must be made in relation to costs of use rates planned for the future. Of particular importance is the decision of how far to follow a vein that becomes gradually thinner or less pure. A decision to shift production to more promising veins means frequently that ore bodies not extracted before the shift are permanently lost because to maintain accessibility or to construct a new access would be too costly even if prices of ore are expected to rise. Similar problems exist for decisions regarding the treatment of tailings.

If pure competition does not prevail and if the finished products (for example, refined metals) are durable, the mine operator may have to consider the effect of this year's production upon future (marginal) revenues.

Illustrations from the Oil Industry

An oil producer has much sunk investment in wells, pumping systems, pipelines, and storage facilities.

Personal immobility is not an important factor in oil producing.

On the other hand, relations of use rates through competitiveness in costs are very important. Present rates of production affect future rates decisively through their effects upon gas pressure and water intrusion. A rate of pumping of an individual well that is either too fast or too slow increases future costs of production and decreases the ultimate cumulative yield of an oil pool. Similarly, the number of wells drilled into a certain pool and the spacing of these wells (assuming given rates of pumping) have important effects upon gas pressures, costs, and ultimate cumulative yield.

If the whole pool is not under unified control of one operator or a group of operators, the individual operator must regard oil and gas as a fugitive resource, as will be discussed in chapter 10. This means the resource can be "captured"—that is, reduced to legal possession—by whoever gets there first and works fastest. Under such conditions, the production plan of an individual operator has to take account of the production plans of all other operators working on the same pool. The same conditions are important for deciding whether or not to pressurize a well artificially or to use other techniques of increasing future rates of production and ultimate cumulative yield. Under monopolistic conditions, rates of production and storage may have effects upon future revenues.

Although many more examples of the interrelations of use rates in different planning intervals could be given, enough has been said to show that no planning agent can avoid making decisions about conservation. We may now ask: What is the objective of planning agents in making such decisions? In other words, what is the optimum state of conservation in private economics?

CHAPTER

6

THE OPTIMUM STATE OF CONSERVATION

The Optimum State Ex Ante *and* Ex Post

The optimum state of conservation as the objective of decisions by planning agents in private economics is an *ex ante* concept. Can such an objective be formulated more precisely and in a way that criteria for conservation decisions are provided? Can a formulation be found that is useful as a guide for individual planning agents and for understanding (and possibly predicting and influencing) their conservation decisions? What are the major objections against such a formulation? Do such objections impose limitations on the usefulness of the concept in economic analysis? What are these limitations?

After an attempt to answer these questions, we will explore in what sense the optimum state of conservation could be used as an *ex post* concept. It will appear that some of the limitations on the usefulness of the *ex ante* concept do not apply to its counterpart *ex post*.

The Optimum State of Conservation and Maximization

In economics, the objective of decisions by planning agents is usually formulated as maximization of some expected value, quantity, or state of well-being—variously called "profit," "income," "utility," or "satisfaction." For the purpose of this discussion we are not greatly interested in the lengthy controversies about what terms are preferable. In accordance with the concepts of revenues and costs employed in chapter 3, we will use the term "net revenues."

If maximization is applied to conservation decisions, a flow of net revenues extending over time must be maximized and the value of physical assets at the end of the planning period must be considered. This latter value is of special interest in the economics of conservation but may be included formally in the net revenues of the last planning interval.

Net revenues accruing at different planning intervals are reduced to present values through discounting with "the prevailing interest rate." The complexities involved in such a use of an interest rate deserve a more detailed discussion, which will be undertaken in the following chapter. At this point, we may assume that these complexities can be solved.

Thus, we may formulate the optimum state of conservation (as an *ex ante* concept) as that time distribution of use rates that maximizes the present value of the flow of (expected) net revenues.

Theoretical Determination of the Optimum State

In economic theory, the optimum state of conservation can be determined in various ways. Revenue and cost functions and the joint-production approach may be used for this purpose.[1] A similar solution may be used in connection with the production function (instead of revenue and cost functions); or net revenues may be formulated as an integral over a variable period of time,

[1] See the Appendix, section 2. Instead of revenue and cost functions, revenue and cost indifference curves can be used. In time economics, however, this variation of the joint-production approach is not so useful because it is desirable to consider use rates in more than two planning intervals simultaneously. The hypersurfaces necessary in this case with the indifference approach are too cumbersome to handle. The indifference approach in joint production is treated in detail in S. V. Ciriacy-Wantrup, "The Economics of Joint Costs in Agriculture," *Journal of Farm Economics*, Vol. 23, No. 4, November 1941, pp. 771–818.

and the calculus of variations employed for maximization.[2]

It was mentioned in chapter 3 that the length of the planning period is, in principle, itself an unknown in the theoretical determination of the optimum state of conservation. In the joint-production approach, this is taken into account by including as variables a number of future rates which in the solution become zero. For each considered change in the planning period, an optimum time pattern of use rates must be determined separately and the results compared in order to determine the optimum state of conservation. In the calculus-of-variations approach, both the optimum length of the planning period and the optimum distribution of use rates are determined by the same operation.

Interest and uncertainty have an important influence upon the extent of the planning period in ways which will be explained in detail in chapters 7 and 8. But interest and uncertainty *by themselves* do not determine the extent of the planning period: they are neither necessary nor sufficient conditions for its end.

Theoretical Determination and Practical Approximation

Theoretical determination of the optimum state of conservation is of interest because criteria for its realization in the sense of "necessary and sufficient conditions" are revealed. However, these criteria are formal. They are not of direct use to planning agents in reality. For them, practical approximations to the formal criteria of economic theory must be found.

The various ways of determining the optimum state of conservation in economic theory, which were just mentioned, are not equally suitable for practical approximations. The joint-production approach has some advantages in this respect. The discontinuity in the method of determining the optimum distribution of use rates over time and the optimum planning period (required in the joint-production approach) renders this method susceptible to practical approximations. The theoretical elegance of the calculus-of-variations approach is of little practical value

[2] These two solutions are discussed in S. V. Ciriacy-Wantrup, "Private Enterprise and Conservation," *Journal of Farm Economics*, Vol. 24, No. 1, February 1942, pp. 75–96.

because this method is less susceptible to such approximations. For this reason only the joint-production approach to the optimum state of conservation is presented in the Appendix.

The relative theoretical clumsiness of the joint-production approach is a smaller disadvantage than may appear at first glance: First, an economically relevant [3] maximum planning period is frequently given. Second, a planning interval may be so defined that the total number of intervals becomes relatively small. Third, as an extreme case of the foregoing simplification, the utilization plan may pass into a stationary state from a certain number of intervals onward. Let us illustrate these simplifications by examples.

A farmer may decide to retire from farming at a certain age. The retirement date would then terminate his planning period.

A tenant may plan to move to another farm when his lease expires. For him the optimum planning period could not be longer than the duration of his lease.

If a time discount is used and if net revenues considered do not increase too rapidly with time, additions to the present value of the series become small beyond a certain distance.[4] A somewhat arbitrary maximum number of intervals can be chosen without materially reducing the highest present value, but sometimes with considerable savings in calculation. If a time discount of, let us say, 4 per cent is used and the series is fairly constant over time, the maximum period chosen need not be much longer than fifty years—although the theoretically optimum period may be longer. In calculations of economic feasibility in public water-resources development, the Corps of Army Engineers employs a maximum period of fifty years and the Bureau of Reclamation, of sixty years —using 3 per cent at present.

If allowance for uncertainty is made through discounting or

---

[3] "Economically relevant" means that the given maximum number of intervals is smaller than the optimum number of intervals.

[4] In these calculations, revenues and costs are dated for those intervals in which they actually occur. This holds also for the initial installation, the maintenance, the salvage value, and the replacement of durable cost factors. Amortization of initial costs—needed for satisfying tax requirements and other purposes of accounting—is neither necessary nor helpful for the present purpose.

flexibility or both, it is frequently reasonable in private enterprises to choose a considerably smaller maximum number of intervals than would be chosen because of interest. On the other hand, allowance for uncertainty may be made in such ways that the problem of choosing a finite planning period is eliminated. These aspects of allowing for uncertainty will be more fully discussed in chapters 8 and 18.

In agriculture, a whole crop rotation may be taken as one planning interval for the purpose of production planning. Hence, even if a farmer intends to plan his production for his whole active economic life, not more than six to ten intervals are to be considered if he uses one of the more common four-year rotations, like the Norfolk rotation.

In forestry, a whole cutting cycle from twenty years or less for pulpwood to a hundred years or more for hardwood timber may be defined as one planning interval.

The last interval of the planning period may be defined as a stationary state of indefinite length in terms of clock time. Sometimes this simplification is so used that the maximization problem over time becomes identical with the instantaneous maximization problem. This occurs if planning intervals previous to the stationary state are disregarded. For example, conservation of flow resources in agriculture and forestry is often regarded as a change from one stationary state to another.

As this particular case indicates, one should keep in mind how practical approximations of the optimum state of conservation are related to the formal solutions. Approximations can be successfully applied only if their weaknesses are understood. Let us now focus on these approximations.

Approximation Step by Step

In economic reality, it is usually not practical to make, or to plan, many small changes of variables (use rates, revenues, costs, net revenues) as theoretically required for maximization. Only a few "lumpy" changes of irregular magnitude can be considered. Furthermore, a change in net revenues must be large enough to be an incentive for action by planning agents. In conservation decisions, as in any other form of behavior, a stimulus must reach a certain absolute size (threshold) to elicit a response.

Under such conditions it becomes impractical to determine the optimum state of conservation accurately. The practical objective of conservation decisions is not the optimum distribution of use rates over time, but a step-by-step directional (conservation or depletion) change of the existing or some hypothetical distribution toward the optimum. The optimum state of conservation can only be approximated step by step, through trial and error. Very small (marginal) changes in use rates, revenues, costs, and net revenues employed in economic theory are replaced by discrete changes. Partial derivatives are replaced by total additional quantities. An increment of present net revenues rather than maximum present net revenues is the objective and criterion of conservation decisions.

The step-by-step approach to the optimum state of conservation has certain hazards if there are a number of economic optima, one of which is the *optimum optimorum.* Conceivably, planning agents using the step-by-step approach may get stuck on a certain peak of net revenues—that is, may not discover the highest peak. This depends on the boldness with which new trials are undertaken. It is incorrect, however, to contend—as some economists have done—that a step-by-step approach is inherently incapable of dealing with a multiple-peak situation. Such contentions stem from the illusions concerning the normality of continuous net-revenue curves still fostered by most economic textbooks.

A Further Simplification

Although a discontinuous, step-by-step approach simplifies considerably the task of approximating the optimum state of conservation, there is a further simplification which is significant, both in the private and the social economics of conservation. This simplification is based on the concept of the production function.

Use of revenue and cost functions is most convenient in economic theory. In economic reality, planning agents are inclined to think in terms which the economist would describe as the "production function."

The physical inputs of (dated) productive services, their technical relation to outputs (the time pattern of use rates), and the technical relations among the productive services themselves—they need not be independent—are given by the production function and by the side relations to which this function may be subject.[5]

Consideration of the production function is *implied* in the theoretical determination of the optimum state of conservation through revenue and cost functions: for each time distribution of use rates that is considered, productive services have to be combined in such a way that the present value of total costs is a minimum. Equiproportional changes of productive services with changes in the time distribution of use rates are rare. *Consideration of the production function is, therefore, always necessary when such changes are contemplated.*

If the step-by-step approach through lumpy changes is applied to the data given by the production function, the concept of a "conservation practice" is obtained.

Conservation Practices

When planning agents make conservation decisions, they usually consider definite combinations, quanta, of interrelated productive services. Such combinations will here be called "conservation practices." A given conservation practice involves interrelated changes in many productive services in more than one planning interval; likewise, products (use rates) may be changed in many intervals.

For example, in soil conservation, the practice of terracing involves inputs of man-hours and machine-hours for construction, the expenditure of materials for preparing proper outlets, repair and maintenance over the years, variations in size and layout of fields, changes in methods of cultivating and harvesting, changes in yields, changes in risks, and possibly other changes.

Under such conditions, the (present) total additional value

[5] For a more detailed discussion of the production function in joint production, see the article cited in footnote 1.

product of the conservation practice is compared with its (present) total additional costs. Usually, the profitability of only a small number of alternative conservation practices is calculated (budgeted).

Although no one can accurately determine the optimum state of conservation in this way, save by accident, the modest but practical goal of approximating the optimum step by step can be effectively pursued.

Objections and Limitations

The foregoing approximations make the optimum state of conservation as an *ex ante* concept more useful for individual planning agents and for the observer who wants to understand, predict, or influence conservation decisions of planning agents on the basis of their motives, objectives, and criteria.

For such an observer—that is, in this study—the optimum state of conservation is a construct, an intervening variable, which is designed to help in understanding the *direction of change* in the time distribution of resource use (conservation or depletion) and in explaining the *modus operandi* of the forces which bring such a change about. The optimum state of conservation, as the concept is used here, does not imply that such a state actually is or could ever be realized in a world of ceaseless change.

There are a number of objections which may be raised against the usefulness of our construct. Some objections do not appear valid; others, however, limit its usefulness. These objections and limitations will now be taken up under four headings: irrationality, extramarket values, uncertainty, and habit patterns.

Irrationality

The usefulness of the optimum state of conservation as an *ex ante* concept may be denied by pointing to "irrationality" in conservation decisions. This term may have a number of different meanings.

The term "irrational" may mean that conservation decisions are made which are not meaningful in terms of a *subjective* appraisal of the environment, or, for short, which do not reflect learning.[6] But everyday observation and scientific experiments with humans and animals show that, normally, choices are made which are meaningful in terms of a subjective appraisal of the environment and which reflect learning. Some conservation decisions, to be sure, are made impulsively and contrary to learning. Psychological analysis may suggest the subconscious rationality of such decisions. In economics they may be regarded as irrational, but as exceptional cases not relevant for most purposes of analysis. The term "irrational" will be understood here in this sense.

The term "irrational" may mean that conservation decisions are wrong or inadequate in terms of an *objective* appraisal of the environment (that is, by somebody other than the planning agent—the observer). As we know (chapter 3), it is always the subjective appraisal by the planning agent himself which is the basis of an ends-means scheme. The term "irrational" will not be used here in this sense.

This subjective nature of the optimum state of conservation would interfere with its usefulness if there were no regularities in environmental appraisal, objectives, and criteria for each individual planning agent, and if homogeneous groups of planning agents could not be found. It is common knowledge that such regularities and groups exist. In the economics of conservation, it is often permissible to reason in terms of a "normal" appraisal of the environment and "normal" behavior by planning agents on the basis of such an appraisal. The nature of the dispersion around the norm can be ascertained through appropriate empirical studies.

By some authors, conservation decisions are called "irrational" if they do not follow that particular norm of behavior they assume for *homo economicus* in modern capitalistic societies. Such an arbitrary use of the term appears misleading. There are important types of behavior that deviate from that of the *homo economicus* but are by no means contrary to learning. Such "deviations" will be discussed presently.

---

[6] "Avoidance of logical and arithmetical errors"—sometimes used as a criterion of rationality—is a result of learning.

Extramarket Values

There is no need to labor the point that conservation decisions are not solely influenced by positive and negative values which are expressed in monetary terms—that is, which are valuated through and in the market place. The availability of leisure time and good working relations with family members, neighbors, employees, and government officials are common nonmonetary considerations in production planning. Prestige, power, pride in his own achievement, aesthetic appreciation of a well-kept farm or landscape may be as important for a planning agent as monetary rewards. To remain in exclusive control of one's business may be regarded as more essential than to increase pecuniary net revenues. Group-centered motives, such as patriotism and community spirit, or envy and vengefulness may influence conservation decisions.

These values are usually called "intangible" in economic analysis. The implication of this term appears unfortunate. In this study they will be called "extramarket values."

For individual planning agents, inclusion of extramarket values in revenues and costs does not prevent approximating the optimum state of conservation. It can be established through observation of behavior, through interrogation, and through introspection that planning agents are able to compare changes in their state of well-being connected with *ex ante* changes in the combination of extramarket and market goods. Thus, they are able to take account of both types of goods in their conservation decisions. The psychological mechanism of these subjective evaluations themselves (for example, whether cardinal or ordinal differentiation of utility is involved) are neither accessible nor relevant for the observer—that is, for *objective* evaluation of extramarket goods.

Objective evaluation of extramarket goods can sometimes be accomplished by analogy (by using market values in auxiliary calculations), or by employing market criteria—such as equality of supply and demand. Sometimes rates of substitution [7] be-

[7] These rates are marginal rates for a given level of indifference. Practical application will be discussed later (chap. 17).

tween extramarket and market goods can be obtained objectively through observation of the planning agent's behavior in situations of choosing, either actual or hypothetical (in questionnaires, for example). Objective evaluation by these various methods makes the optimum state of conservation an operationally useful concept for public policy in spite of the existence of extramarket goods. More will be said about the valuation of extramarket goods when the objectives and criteria of public conservation policy are discussed in chapter 17.

The foregoing carries no implication that conservation decisions could not be studied partially—that is, for market values alone, by keeping extramarket values constant; or vice versa. It should be realized, however, that market values and extramarket values are frequently interrelated. To keep constant means to take account of these relations.

We may conclude that the two objections against our construct discussed so far—based on the existence of irrationality and extramarket values—do not seriously limit the usefulness of the optimum state of conservation as an *ex ante* concept. Let us see whether this holds also for the two objections to be discussed next. These objections are based on the existence of uncertainty and of habit patterns.

Uncertainty

It is commonly recognized that planning agents frequently have only hunches and broad guesses with respect to the economic data necessary for determining the optimum state of conservation. Expectations are uncertain. In more technical language, expectations are not single-valued but appear in the form of a probability distribution. Planning agents seldom have exact numerical knowledge of the statistical moments of this distribution. But they have, generally, some notion about the range of possible outcomes and, sometimes, about the most probable outcome. The question arises whether and in what way this situation influences the objectives and criteria of conservation decisions.

On one side it can be argued that there is no such influence:

planning agents employ in their calculations the most probable value of expected net revenues and discount this value for uncertainty—that is, reduce it in proportion to dispersion. Thus, multivalued expectations are treated in production planning as if they were single-valued.

On the other side it can be pointed out that uncertainty of expectations cannot effectively be taken into account in this way. It may be conceded that, as a matter of business practice, planning agents sometimes take uncertainty into account by using the most probable value and discounting for dispersion. Without disputing the actual extent of this practice, it has two serious limitations:

First, discounting is ineffective if the most probable value of net revenues cannot be ascertained. Such situations are very common in economic reality because, as just emphasized, expectations frequently consist merely in hunches and vague notions of the range of possible occurrences.

Second, discounting may be an ineffective allowance for uncertainty even if the most probable value of net revenues can be ascertained. This is true if the possibility of a highly unfavorable outcome exists—even though such an outcome may be much less probable than alternative more favorable outcomes. Discounting the most probable net revenues for the possibility of high negative net revenues (losses) may not lead to decisions that would safeguard the enterprise against bankruptcy if the most unfavorable outcome is actually realized. This threat depends both on the magnitude of the loss and on the financial strength of the planning agent—which always has definite limits.

A contingency which threatens bankruptcy can be guarded against in various ways which will be discussed in detail later (chapter 8)—for example, through hedging, pooling, and spreading arrangements, or by avoiding commitments which would cause dangerous losses if the most unfavorable outcome should occur—that is, by keeping the production plan flexible. Flexibility is a way of allowing for uncertainty of expectations because most uncertainties increase with distance in time: planning agents know that expectation about a certain planning interval will become less uncertain the nearer this interval is approached in the course of time.

Avoiding the possibility of the most unfavorable outcome involves either definite costs—for example, a risk premium for an insurance policy—or possible losses through flexibility. Losses occur with a more flexible plan—as compared with a less flexible plan that would bring maximum net revenues under the most probable expectation—if the most probable outcome is actually realized or, at least, if the most unfavorable outcome is not realized. Thus, the more flexible plan may not be the optimum plan as defined through maximizing the most probable net revenues; this, indeed, is highly probable. However, these more probable losses through greater flexibility are moderate as compared with the losses through smaller flexibility *if* the most unfavorable outcome is realized.

In other words, one important objective of conservation decisions is to avoid *immoderate* possible losses—although of small probability—by accepting the possibility of moderate ones—although the latter are more probable. For our purposes, a loss may be called immoderate if it threatens the continuity of a production plan.

Economic choices between losses of various magnitudes and probabilities exist not only in those special cases in which immoderate losses, in the sense just indicated, are under consideration. Such cases were mentioned first because they reveal clearly the insufficiency of taking account of uncertainty through discounting. Much more numerous are cases in which alternatives exist between larger but less probable losses and smaller but more probable ones—although the former need not be immoderate.

It may be concluded that multivalued decision problems are so common in economics that the objectives and criteria of conservation decisions are best formulated in a way which takes uncertainty explicitly into account. This can be done, for example, by subjecting the economic optimum to the restriction of avoiding immoderate possible losses, or by formulating it as "minimizing maximum possible losses," or, more generally, as a set of rules for choosing among alternative probability distributions.[8] Implicitly, however, all such formulations can still be included in a concept of maximization.

The more specific effects of uncertainty upon the state of conservation will be discussed in chapter 8 for private economics and in chapters 17 and 18 for social economics.

Habit Patterns

Planning agents frequently do not act on the basis of a conscious, calculative comparison of alternatives but on the basis of what will be called here "habit patterns"—tradition, custom, routine, following the leader, drill. Habit patterns are learned through imitation and repetition, not through understanding based on analysis.

The models of economic theory currently in vogue disregard habit patterns. In the economics of conservation, especially of flow resources, the significance of habit patterns is so great that such an omission would lead to an unrealistic picture of resource use. We have to ask, therefore: how is the usefulness of our construct affected by the existence of habit patterns?

There is little doubt that the optimum state of conservation as an *ex ante* concept loses its usefulness for economic analysis if conservation decisions are dominated by habit patterns.

True, by a conceptional *tour de force*, habit patterns may be regarded as a tool of maximization rather than its negation: habit patterns reduce greatly the effort of choosing in familiar situations through diminishing the number of consciously considered alternatives; the planning agent's attention is thus free for decisions when new problems must be faced; without such an economizing in the effort of choosing, the great mass of daily choices could not be coördinated.

However, nothing is gained for economic analysis by including habit patterns in this way as a tool of realizing the optimum. Be-

[8] Formulation of such a set is attempted by Johann von Neumann and Oskar Morgenstern in *Theory of Games and Economic Behavior* (Princeton: Princeton University Press, 1944), 625 pp.

havior is not explained by defining *all* behavior, from whatever nonobservable cause, as optimal behavior. A conceptional structure that is not helpful for explaining behavior must be discarded.

A frankly historical, institutional, and sociological approach to habit patterns would seem more promising than merely defining such behavior as optimal. The economist can make a contribution by analyzing the origin, development, and effects of habit patterns; but a maximization theorem, which is by no means identical with economic theory as such, is of little use in this analysis.

Generally, some conservation decisions are influenced by habit patterns and others by a conscious comparison of economic alternatives. This appears true even for social groups in which habit patterns have a strong influence upon conservation decisions, as in some tribal societies in Africa.[9] In these cases, economic theory based on a maximization theorem is useful if it is employed together with a careful institutional and sociological study of habit patterns.

Economic theory, of course, can operate if habit patterns are kept constant—that is, if their changes are properly accounted for. Over those periods of time with which conservation is concerned, habit patterns may actually not remain unchanged. Such changes are important because economic choices on the basis of habit patterns are related in many ways to economic choices made on the basis of conscious comparison of alternatives.

There is an important special case of habit patterns and of their relation to conscious comparison of alternatives. This special case exists when social mores and institutions favor a calculative, antitraditional, and materialistic (if not outright pecuniary) approach to conservation decisions. Such a case prevails in modern societies based on division of labor and exchange—socialistic as well as capitalistic. This study deals primarily with the economics of conservation in these societies.

---

[9] A. I. Richards, *Land, Labour and Diet in Northern Rhodesia: An Economic Study of the Bemba Tribe* (London: Oxford University Press, 1939), 415 pp.

D. M. Goodfellow, *Principles of Economic Sociology: The Economics of Primitive Life as Illustrated from the Bantu Peoples of South and East Africa* (Philadelphia: P. Blakiston's Son and Co., 1939), 289 pp.

Optimum State of Conservation as Ex Post *Concept*

The foregoing limitations referred to the usefulness of the optimum state of conservation as an *ex ante* concept—that is, as a construct of economic theory to help in explaining the behavior of planning agents through their motives, objectives, and criteria.

We saw earlier that the concepts of "conservation" and "state of conservation" may be used in the *ex post* sense—that is, may refer to the consequences, the outcome, of the actions of planning agents regardless of their motives and objectives. In this sense, observable changes in the state of conservation may be called "toward" and "away" from the optimum in terms of relative economic success (survival) in a given economic environment.

No motives and objectives—such as maximization or any other —*need* be imputed to planning agents. As a theoretical extreme, their actions may be regarded as random. Still, a tendency toward the optimum would exist for a given statistical population of planning agents, because those who are more successful in adapting to environmental forces would tend to survive.

Thus the observer may be able to understand, predict, and influence the aggregate result of conservation decisions by a whole statistical population of planning agents on the basis of his knowledge about the survival value of various practices in a given economic environment. In this case—as in the preceding one—the concept of an optimum state of conservation is useful mainly if the observer is interested in the *direction of change*—as in this study.

Ex Post *Appraisal of Habit Patterns*

The great advantage of such an *ex post* (statistical) interpretation of the optimum state of conservation is that habit patterns can be fully integrated into economic analysis. As we know, the optimum state of conservation as an *ex ante* concept loses its significance for economic analysis if the actions of planning agents are dominated by habit patterns. On the other hand, with the *ex post* concept, habit patterns can be appraised.

Furthermore, habit patterns may be studied as an important

mechanism in the movement toward the optimum. For example, if following the leader (imitation, drill) plays an important role in a given statistical population of planning agents, and if the economically successful ones—rather than those with military, scientific, or religious prestige—are regarded as leaders, successful adaptation to the environment will spread more quickly than it would if only the forces of selection were operative.

Habit patterns do not necessarily favor conservation. True, over long periods, an unchanged environment may select habit patterns which favor continuity of resource use. However, habit patterns well adapted to one kind of environment may become a major liability for continuity of resource use if the environment changes—for example, if habit patterns are transplanted into a different environment through migration or conquest. Resource use in the Old and the New World offers many examples which illustrate resource depletion caused by transplantation of habit patterns. An appraisal of habit patterns is an important aspect of the economics of conservation and a change of habit patterns may be an important objective of conservation policy. We will return to these problems later.

Habit Patterns and Economic Calculation

It has already been indicated that, generally, not all practices in resource use are dominated by habit patterns even in social groups in which habit patterns are strong. It is possible that some practices may be determined by habit patterns, and at the same time others be adopted on the basis of economic calculation.

It is also possible that a state of conservation based on habit patterns may be gradually replaced by an equivalent state based on economic calculations. Such a replacement does not occur for all planning agents in a social group at the same time. Thus, in a social group at a given time we may have the same state of conservation adopted for quite different reasons. A historical example for such a replacement of habit patterns by economic calculation is described elsewhere for soil conservation in Europe at the beginning of the industrial revolution.[10]

[10] S. V. Ciriacy-Wantrup, "Soil Conservation in European Farm Management," *Journal of Farm Economics,* Vol. 20, No. 1, February 1938, pp. 86–101.

The coexistence of habit patterns and of economic calculation both in an individual planning agent and in a group of planning agents is not without significance for a public policy that strives toward adoption of a certain state of conservation by all resource users: when an appeal to economic calculativeness is made, care must be taken that the private optimum thus established as an objective does not represent a lower state of conservation than that which might be in existence on the basis of habit patterns.

In some cases it may be advisable, in the interest of a social optimum, to maintain or to strengthen habit patterns rather than to appeal to economic calculativeness. On the other hand, it is possible that habit patterns must be changed or weakened if the private optimum is a higher state of conservation than the existing one. Thus the coexistence of habit patterns and of economic calculations requires institutional insight and educational skill on the part of those who are responsible for public conservation policy.

Optimum State of Conservation and Economic Forces

This abstract discussion of conservation decisions and of the optimum state of conservation does not explain conservation and depletion in actual situations of resource use. But such a discussion provides the organizing principle by which actual situations may be understood.

To use an analogy, the concepts of "climax type" in ecology and "adaptive peak" in genetics do not explain an actual plant association or the developmental state of a species at a particular time and place; nor do they indicate that static states are realizable or that the system which is being considered is closed. Still they are helpful constructs in understanding the direction of ceaseless change, the resultant of environmental forces which can be observed at a given time and place.

Similarly, the optimum state of conservation both as an *ex ante* and as an *ex post* concept is a construct which is helpful as an organizing principle in analyzing the result of economic forces which influence conservation and depletion. A study of these forces is the central theme of the economics of conservation.

PART III

THE PRIVATE ECONOMICS OF CONSERVATION

ECONOMIC FORCES

CHAPTER 7

INTEREST, TIME PREFERENCE, AND INCOME

Why Are We Interested in Economic Forces?

An economic study of conservation must explain how a state of conservation and its changes come about. The variables involved in such a study may be called "economic forces"—including the economic effects of social institutions.

An understanding of these forces is necessary for the following four reasons: First, for explaining the behavior of resource users in the past. Second, for predicting the behavior of resource users under given assumptions with respect to such forces. Third, for understanding the selective processes that operate among a statistical population of resource users and, over time, mold the state of conservation. Fourth, for designing appropriate conservation policies: economic forces may be obstacles or tools for such policies.

General Effects of Interest Changes on Conservation

Among economic forces affecting conservation, interest and related forces to be discussed in this chapter are among the most powerful, most consistent, and, from the standpoint of theoretical analysis and practical effects, among the most clear-cut. For this reason, interest is discussed first.

Interest rates are used in production planning for making net revenues occurring in different planning intervals comparable in time. Future net revenues are discounted to the present. This means (with positive interest rates),[1] that future net revenues that are numerically identical but occur in different intervals are decreased in relation to their distance in time from that interval in which decisions are made. As with definition and measurement of conservation, this relation to distance is the first power of time (chapter 4). An increase in interest rates means, therefore, a progressive decrease (one that becomes greater with distance) in the present value of future net revenues—the progression being proportional to distance. The result will be an attempt by planning agents to change the time distribution of net revenues in the direction of the present. This can be accomplished through redistributing revenues in the direction of the present or through redistributing costs in the direction of the future or through both.

Except for the relatively minor possibilities of reducing storage, revenues are redistributed in this way through redistributing rates of use in the same direction (that is, toward the present).

Costs are redistributed in this way by substituting productive services with shorter periods of gestation for those with longer periods or, in the terminology used in chapter 5, by reducing the periods over which costs are sunk.[2] This also means a redistribution of rates of use in the direction of the present if the state of technology is assumed to be unchanged.

Thus an increase of interest rates tends to change the distribution of use rates in the direction of the present. According to our definitions, this means depletion. By the same reasoning, a decrease of interest rates leads to conservation.

---

[1] Existence of negative interest rates has little practical significance. Theoretically, their influence can be treated in the same fashion as that of positive rates.

[2] As a kind of shorthand, we may speak of "reducing sunk costs," although the *amount* of sunk costs is not necessarily reduced. There is no need to talk about an *average* period of gestation (sunkness); the economic forces discussed here affect periods of gestation of all types of sunk costs in the same direction (chap. 5).

Thus far, we have considered the effects of interest rates upon conservation decisions of individual planning agents. That means we have considered conservation and depletion *ex ante*. However, the effects of interest rates just discussed are not necessarily dependent on the assumption of a calculating, decision-making agent: A rise of interest rates not only makes those planning agents relatively more successful who respond to this change of environment by depleting their resources, but also those who for any other reasons or for no particular reason (at random) have a lower state of conservation than others. Thus, over time, selection and imitation (chapter 6) would bring about depletion *ex post* considering a population of resource users as a whole. Similarly, conservation *ex post* would be brought about by a fall of interest rates.

Interest and the End of the Planning Period

As was explained earlier, in production planning the extent of the planning period is implicit in determining the optimum state of conservation. If an economically relevant maximum planning period is not given—it frequently is given in economic reality (chapter 6)—the extent of the planning period is a variable (both in terms of intervals it includes and in terms of clock time). It does not pay, from a certain interval onward, to obtain net revenues in any additional interval at the sacrifice of net revenues in preceding intervals. Thus the planning period is terminated.

The interrelations of different intervals in net revenues (revenues and costs) are affected by many factors already mentioned. Usually, the extent of the planning period would be a finite magnitude (at least in terms of intervals) even if no interest existed. Existence of interest is neither a necessary nor a sufficient condition for bringing the planning period to an end. Still, interest is a very important economic force in this connection. Interest decreases progressively (i.e., its effect increases with distance in time) the influence of future net revenues upon the shape and extent of the production plan. In other words, interest helps to terminate (at least in terms of intervals) the planning period sooner than would be the case otherwise. This is true also for the clock-time extent of the planning period if the production plan does not lead to a stationary state. A stationary state is compatible with the existence of interest. As will be shown in the next chapter, a somewhat similar situation prevails with uncertainty.

Do Planning Agents Use Market Interest Rates?

These important, but analytically rather simple, general effects of changes of interest rates upon the state of conservation are not the only problems that are significant here. For conservation and depletion *ex ante,* it is essential to know how "internal" interest rates (those used in private utilization plans) are determined.

At first glance, such internal rates are identical with market rates—that is, they are prices determined in and through the market place. Interest rates in this sense balance the current market supply and demand for money funds loanable over various periods of time (long-term bonds, intermediate loans, and short-term credit). The many, sometimes rather complex and controversial, forces which determine the current supply and demand for money are of no direct concern for our present purposes. In the private economics of conservation—in contrast to the social economics of conservation—we are interested in whether and how far market rates are relevant for the utilization plans of individual resource users.

Market rates would be relevant for individual resource users if they could easily exchange on the market their expected future net revenues discounted on the basis of such rates. In terms more familiar to the economist, we may say that market rates would be relevant for individual planning agents if the markets for their "assets" were perfect.[3] In economics, a "perfect market" is defined similarly to "pure competition" (chapter 5)—namely, as a market in which all individuals can purchase and sell any desired quantities at the prevailing price; the spread between asking and

[3] Assets "yield" future net revenues. Assets themselves refer to the present. Planning agents may hold positive and negative assets in three forms: (1) physical assets (natural resources, improvements, equipment, inventories); (2) money assets (securities, loans, cash); (3) personal assets (skills of the planning agent and his family).

selling prices is small and determined by the costs of marketing. A market may be imperfect because it involves few buyers, or few sellers, or both; such a market may be called impure or monopolistic; or a market may be imperfect for other reasons, for example, imperfect knowledge of buyers or sellers about market conditions, or imperfect fungibility of products or services.

Under perfect markets for their assets, individual resource users would scarcely use *higher* interest rates in their utilization plans than those established through the market. Instead of planning (on the basis of higher internal rates) a time distribution of use rates with a lower state of conservation than they would plan on the basis of market rates, they would do better to sell or lease their physical assets and seek employment elsewhere; or, if they want to stay in their business, they would do better to borrow money instead of reducing the market value of their assets by adopting a lower state of conservation.

Likewise, under perfect markets for their assets, individual resource users would not use internal interest rates *lower* than market rates in their utilization plans. Instead of planning (on the basis of lower internal rates) a state of conservation higher than that justified by market rates, they would do better to use the higher current net revenues (resulting from the lower state of conservation based on market rates) for buying or renting more physical assets or lending money.

Imperfect Markets for Assets

The preceding section showed why, in analyzing the effects of interest rates upon conservation decisions, it is important to know whether or not markets for assets of individual resource users are perfect.

In the utilization of stock resources (oil, gas, coal, metal ores) large joint-stock companies predominate. The assets of these firms are easily exchanged in the form of shares, and they have ready access to the loan market in its various stratifications (long-term bonds, intermediate loans, and short-term credit). These markets may not be perfect for all firms in all phases of the business cycle; but it would seem fairly realistic to regard them as such for our purpose, and to expect that internal interest rates should correspond by and large to market rates.

The situation is quite different in the utilization of flow resources, particularly in agriculture, grazing, and forestry. First, assets like farms, ranches, and forests, together with their improvements and equipment, can usually be sold and bought only as wholes. Cases in which assets are freely divisible—for instance, in which shares can be bought and sold in the stock market—are of small importance. Thus, because of indivisibility alone, fungibility is rather rare. Second, the most important assets are immobile; strictly speaking, differences in location always exclude fungibility. Third, with a few exceptions—such as the farms of uniform soil and climate in some sections of the Corn Belt—assets have many special technological features that would exclude fungibility even if divisibility and mobility were greater. Thus, buying and selling and renting and leasing of physical assets in agriculture, grazing, and forestry are complex, highly individualistic, and rather important transactions, which cannot be easily and frequently undertaken by planning agents.

The imperfection of the market for physical assets in the utilization of flow resources is strengthened by imperfections of the market for the planning agent's and his family's labor, his occupational skill, or his managerial ability. Exchange of physical assets requires not only adaptation in the occupational life of the planning agent and his family, but also, usually, a considerable adjustment in other spheres. This is closely related to imperfect fungibility of physical assets. Their spatial extent and locational immobility require mobility of the planning agent and his family when physical assets are exchanged. This necessitates, in social relations, many changes that are connected only indirectly with economic activity. A good example is the transfer of assets in agriculture, where the economic characteristics of physical assets are closely connected with the occupational and managerial skill of the planning agent, his family's labor capacity, and his whole mode of life.

Similar imperfections exist in the market for loans. A loan

contract in the field of flow resources is, like an exchange of assets, usually a highly individualistic transaction. Both the amount of the loan and its price (the interest rate) are closely related to the borrower's assets; loans are generally not extended beyond a certain ratio to the value of those assets. Besides limiting the amount of the loan, the value of assets influences interest rates because these rates increase generally with the ratio of loan to asset; the reason is an allowance for costs and uncertainty on the part of the lender. The individualistic nature of the loan contract, both with respect to the amount of the loan and to its price, tends to break up the loan market into segments. Within these segments actual lenders (in contrast to savers) are few and well organized in comparison with borrowers.

The imperfection of the loan market, caused by imperfect fungibility of the loan contract and by monopolistic conditions, makes it possible to discriminate between borrowers not only through differentiation of interest rates but also through rationing the amount of the loan. Rationing is accomplished by varying the maximum ratio of loan to asset, or by varying the standards that govern the appraisal of assets, or by stipulating that a proportion of assets must be in a certain form.

The Individual Time-Preference Rate

In view of these imperfections, the spread between rates relevant for investment and disinvestment is frequently wide, and variable with time, place, and individual. It is often more economical for planning agents to disinvest in resources through depletion than through sale or borrowing, if there are no ready markets in which assets they want to sell are capitalized, or in which funds can be borrowed, at interest rates lower than their individual rates. Conversely, they may find it more economical to invest in resources through conservation than through purchase or lending if there are no ready markets in which assets they want to purchase are capitalized, or in which funds can be lent, at interest rates higher than their individual rates. We may say in such instances that planning agents take their own "time-preference rates," rather than market interest rates, as a basis for conservation decisions.[4]

Individual time-preference rates are important for the conservation of flow resources in modern Western societies even during periods of general economic stability. They are still more important if one is considering societies with less well-developed markets, or periods of depression in the business cycle or longer periods of stagnation. The economics of conservation is concerned with all these conditions.

The concept of an individual time preference is more realistic and analytically more useful than a mere enumeration of some individual immediate needs which may cause differences among resource users with respect to the most desirable time distribution of net revenues. Time preference can be precisely defined; it is operative in terms of subjective evaluation, but it can be objectively measured in the same sense as other extramarket values —that is, in terms of observable rates of substitution.[5]

Although time preference can be precisely defined for purposes of economic analysis, this does not mean that, in economic actuality, individual planning agents always employ numerically precise rates in discounting for time. Actually such discounting is frequently done on the basis of hunches. It makes little difference for our purpose whether the time discount is precise or vague. It is merely necessary that the weight of net-revenue expectations in influencing production plans is decreased in relation to distance in time.

Time-Preference Rates in Relation to Income Levels

The concept of individual time preference is a marginal one (footnote 5). Time-preference rates, therefore, are affected

[4] Usually, but not necessarily, time preference, like interest, is positive—that is, expected values are decreased in proportion to their distance in time.

[5] The "rate of individual time preference" is defined as a ratio between the present marginal utility of pecuniary net revenues or, for short, "money," in more distant future intervals and the present marginal utility of the same amount of money in intervals nearer to the present and reduced by unity; it may be expressed in per cent and per unit of time. Only ratios of marginal utilities are under discussion, and the foregoing definition can be expressed in terms that have no reference to utility: the "rate of individual time preference" may be defined as the numerical gradient of a given indifference level between money in two different intervals, reduced by unity.

by the numerical value of net revenues—or, better for the present purpose, the level of income [6]—of each planning agent. Thus, in the economics of conservation we are greatly interested in the way changes of income levels affect time-preference rates.

On the basis of observation of human behavior, economists generally accept the theorem that a unit of income—a dollar—becomes less and less effective in influencing economic decisions as income increases.[7] This theorem applies to instantaneous economics or, in our terminology, to each planning interval. In time economics, we are interested in the ratio between the effectiveness (in influencing economic decisions) of the identical amount of income in different planning intervals. If the theorem is approximated, this ratio must also decrease progressively with increasing income levels.[8] This ratio is identical by definition (footnote 5) with the rate of time preference.

We see readily, therefore, that changes of income levels of planning agents may have profound effects upon their time-preference rates and thereby, if markets for assets and loans are imperfect, upon the state of conservation. Under this assumption, a decrease in income levels will increase time-preference rates and lead to depletion; an increase in income levels will decrease time-preference rates and lead to conservation.

A given change in income level, however, will have less and less effect upon the state of conservation as income increases until these effects disappear altogether.[9] In other words, the relations between income level and conservation decisions are most im-

---

[6] For the present purpose "net revenues" must include the earnings of the planning agent himself and, sometimes, of his family. For other purposes of analysis these items may be regarded as costs. Likewise, for our present purpose we must include earnings from other activities than resource utilization.

[7] More precisely, in instantaneous economics, the marginal utility of money is assumed to decrease with increasing income and to approach the abscissa (on which income is plotted) asymptotically. There are some doubts whether this decrease is always monotonical. However, the general character of the relation and the conclusions based on it (see following footnote) are not invalidated by these doubts.

[8] Under the assumptions stated in the preceding footnote, ratios between corresponding points on schedules of present marginal utility of income in different planning intervals must be the largest at the lowest income and must decrease with increasing income until these ratios become very small.

[9] This also follows from the functional relation between income and marginal utility of money assumed here and stated in footnotes 5 and 7.

portant in the low-income groups of resource users, and, for all groups, have more importance during depressions than during booms.

Regression, Progression, and Fixity

From the relations discussed in the above section, it follows that, in conservation economics, we are interested in those income-changing forces which affect different income levels differently or, in terms borrowed from the economics of taxation which are "regressive" or "progressive" with income, and those forces which do not vary with income during general income changes—as during depressions. The latter forces are commonly called "fixed charges." [10]

Under the assumption of imperfect markets for assets, regression of those forces that *decrease* income tends toward depletion as compared with a situation in which the same aggregate decrease (that is, for all income levels considered) takes place without regression. Conversely, regression of those forces that *increase* income tends toward conservation as compared with a situation without regression. Progression of income-changing forces has the opposite effects upon the state of conservation. Let us illustrate the problem of regression and progression with some examples. For this purpose we will interpret "income-changing forces" to include prices of products and of productive services, subsidies, and charges. Regression in subsidies and product prices tends toward conservation, regression in charges and prices of productive services toward depletion.

Prices of products are frequently progressive: they vary with scale of operations; income of planning agents varies likewise with scale of operations. It is often more expensive per unit to

[10] Fixed charges are different from fixed costs mentioned in chapter 5. Fixed charges are amounts of money (cash) which planning agents are obligated to pay in each planning interval regardless of net revenues or income obtained. Fixed charges are often, but not necessarily, incurred in securing those factors with which fixed costs are connected (payments of cash land rent, property taxes, interest payments for loans to purchase durable goods). However, fixed charges may be incurred in connection with securing variable services (interest payment for feed and fertilizer loans). Finally, fixed charges may not be costs at all (interest payments on a personal loan or cash rent on a home).

market a smaller volume than a larger one. Small producers have less complete information about market conditions. They are more susceptible to monopsonistic exploitation. They are less able to practice monopolistic exploitation. Finally, the ratio of products marketed to products consumed at home varies with scale of operations. For all these reasons price increases—for example, through government support—are generally less effective in bringing about conservation than if the same aggregate increase of income is accomplished through direct income subsidies. Furthermore, such income subsidies can deliberately be made regressive by the government.

Prices of some productive services are regressive for reasons similar to those that explain progressive prices of products (greater expenses of purchasing a smaller volume, inferior market information of small producers, monopolistic exploitation of small producers, inability of small producers to practice monopsonistic exploitation). However, such regression rarely exists in the prices of hired labor; and below a certain scale of operations, labor furnished by the operator and his family reduces the dependence on hired labor. Regressiveness in prices of productive services, therefore, is generally of no great significance in the economics of conservation.

The situation is quite different for charges. Although progression is provided for in some taxes (for example, on personal income and profits), others are (unintentionally) regressive (chapter 13). Debtors in low-income groups pay sometimes higher interest rates than debtors with large incomes (chapter 12); with the former group, risks are considered greater by some lenders, and the costs of administering smaller loans are higher per unit. Rents are often regressive because the demand for small properties is greater than the demand for large properties—after allowances are made for the differences in income-yielding capacity (chapter 11); the causes are limitations of managerial capacity, immobility of operators, and imperfections of the loan market.

Let us now turn to the problem of fixity. Under the assumption of imperfect markets for assets, fixed charges tend toward depletion as compared with a situation in which the same amount is paid during a whole business cycle (that is, including a boom and a depression) in the form of flexible charges. Depletion occurs during depressions. This depletion is not materially offset through conservation during booms. During booms, incomes are higher than during depressions. As we know, changes of incomes (in this case a comparatively smaller reduction of income through fixed charges) have less and less influence upon time-preference rates as incomes increase.

From the standpoint of resource utilization, the effects of fixed charges are undesirable both in private and in social economics. Their alleged advantages from other standpoints—for example, in giving stability to tax yields and to the income of landlords and creditors and thus maintaining community services and attracting outside capital during depressions—are largely illusory. What happened to this "stability" during the "great" depression in the beginning of the 1930's is common knowledge. This happened in spite of large-scale financial assistance by the federal government. Similar experiences are recorded repeatedly for all major depressions in this and in other countries.

On the other hand, it is practical to transform most fixed charges into flexible ones. Property taxes can be transformed into income and severance taxes; fixed cash rents can be transformed into flexible cash rents or share rents; in some countries successful experiments have been made in expressing loans and interest payments in agriculture in terms of the main products—corn, wheat, cotton—or in terms of units of a composite of crops and livestock. More will be said about these possibilities when the effects of taxation, tenure, and credit upon conservation are discussed.

Effects Accentuated in Underdeveloped Countries

The relations between conservation decisions, interest rates, time-preference rates, and income, which were discussed in the preceding sections, are important for the most developed industrial countries with high incomes and capital formation (the United States, western Europe, British Commonwealth). In these countries they are particularly significant for flow resources, within the low-income groups of planning agents, and during depressions. They are of even broader significance for industrially underdeveloped countries with great demand for but small supply of loanable funds. In these, interest rates are high, imperfections of markets for assets are numerous and effective, incomes are low, and fixed charges (particularly through monetization of increasing taxes and indebtedness and substitution of cash rents for share rents) are increasing. In the absence of counterbalancing forces, therefore, one would expect rapid resource depletion.

In some of these countries such counterbalancing forces were present until the nineteenth century through habit patterns which safeguarded a "minimum standard of conservation" (chapter 18) in the utilization of land, forest, water, and wildlife. During the nineteenth and twentieth centuries these habit patterns have been gradually weakened by the economic, institutional, and psychological impact of Western industrial civilization. Since then the economic forces discussed in this and in the following chapters have attained greater and greater significance for resource utilization in underdeveloped countries. The result has been widespread and rapid resource depletion.

Implications for Public Conservation Policies

The economic forces discussed so far have been and are important causes for resource depletion. On the other hand, an understanding of them offers great opportunities for public conservation policies. Such opportunities (through changing interest rates, time-preference rates, incomes, and fixed charges, and correcting imperfections in the markets for assets) are only implied in this chapter. They will be more fully considered in the chapters on taxation, tenure, and credit, and in Part IV, which deals with the objectives and criteria of public conservation policy.

It has already become evident, however, that a merely technological or educational approach in conservation policy may be of little avail if the economic forces affecting conservation and depletion are imperfectly understood. Natural resources are frequently used wastefully (from the standpoint of private economics, or social economics, or both) not because individual resource users do not know any better, but because they cannot help it under the influence of economic forces. If this cold fact were better understood by many well-meaning writers, progress in the conservation field might be more rapid. This holds particularly for industrially underdeveloped countries. Better knowledge—from contour farming to birth control—may be of little help if the influences of high interest rates, imperfect markets for assets, low incomes, high fixed charges, and deterioration of habit patterns continue unabated.

CHAPTER 8

UNCERTAINTY

Uncertainty and Uncertainty Preference

As we saw in chapter 6, existence of uncertainty refers to the fact that expectations of future revenues and costs are nearly always considered by planning agents to have probabilities of less than unity. That means the expectations are not single-valued but appear in the form of a probability distribution. Such a distribution is essentially subjective (hypothetical), although it is, of course, based in various degrees on experience.

Theoretically, a planning agent would have to know three (or more) [1] characteristics of the probability distribution: the most probable expectation (in terms of mean, mode, or actuarial value), the dispersion around the most probable value (in terms of range or coefficient of variation), and the asymmetry of the dispersion (in terms of Pearson's or Bowley's measure of skewness). In actuality, planning agents have only estimates and hunches about these characteristics.[2] These estimates are gener-

[1] Theoretically, the analysis could be extended to further characteristics (parameters); for example, to any number of moments.

[2] Knight's differentiation between "risk" and "uncertainty" is based on the difference between statistical and estimate probability distributions. See Frank H. Knight, *Risk, Uncertainty, and Profit* (New York: Houghton, Mifflin Co., 1921), 381 pp.

This difference is one of degree. In our study, accordingly, the term "uncertainty" will include both statistical and estimate probabilities.

ally sufficient only to establish a notion about the range of possible outcomes and, sometimes, about the most probable outcome.

Planning agents are especially interested in the range—the extreme values—of the dispersion. The reason was indicated in chapter 6: the extreme values frequently mean great losses or, alternatively, great gains. The majority of planning agents want to avoid immoderate losses, losses that threaten the continuity of the utilization plan. A minority of planning agents, the gamblers, on the other hand, might strive for great gains in spite of and, sometimes, because of the possibility of great losses. Both groups, therefore, will focus on the range of dispersion. Thus, it would seem realistic to interpret uncertainty mainly in terms of the range of dispersion—rather than in terms of skewness or in terms of higher statistical moments of the probability distribution.

As just implied, planning agents have greatly different but generally negative uncertainty preferences.[3] Among gamblers positive uncertainty preferences may occur; but with some possible exceptions (oil drilling) such cases are not important in resource utilization. A negative uncertainty preference means that of two revenue, cost, or net-revenue expectations, with the same dating and with the same most probable value, the less uncertain one—that is, the one with the smaller dispersion—is preferred.[4]

Allowance for Uncertainty

Allowance for uncertainty is affected both by uncertainty of expectations and uncertainty preference. We observed in chapter 6 that planning agents may allow for uncertainty in various ways.

---

[3] The reasons for differences in uncertainty preference are beyond the scope of this study. Age, nationality, community, family, experience, and education—especially during the most impressive years—are some of the factors which need to be considered.

[4] For some purposes of economic theory a more rigorous definition is desirable. We may define individual uncertainty preference in the fashion we defined individual time preference (in chap. 7, footnote 5), with the difference that all values (or "money") are understood as most probable values and are differentiated not on the basis of dating, but on the basis of range of dispersion.

First, they may discount for uncertainty. That means they decrease most probable net revenues that are numerically identical but have different probability distributions in relation to the range of dispersion. Such a decrease does not necessarily require precise uncertainty discounts. It may mean merely that the weight of most probable net revenues in influencing the utilization plan is decreased according to hunches and estimates of the rank (order) with respect to range of dispersion.

Second, planning agents may allow for uncertainty by hedging, pooling, and spreading arrangements. By such arrangements, uncertainty of expected revenues or costs is reduced in production planning. In hedging, the producers of resources, the sellers of futures, shift the incidence of uncertainty to specialists, the professional speculators. In pooling and spreading, there is a decrease in the dispersion of revenues or costs to pool members or to the planning agents (or agent) who participate in spreading.

Third, planning agents may allow for uncertainty by keeping their utilization plan flexible. This means that they may decrease the periods over which costs are sunk, avoid obligations to pay fixed charges (interest, rent, taxes), and accumulate liquid reserves until planning intervals are approached closer in the course of time. Usually, uncertainty of expectations increases with distance in time. This relation between uncertainty and time requires particular emphasis in the economics of conservation.

Uncertainty in Relation to Time

The most important uncertainties—namely, those created by changes of technology, of consumer demand, and of social institutions—increase with time; and uncertainties of nature (drought, floods, insect pests, hailstorms, fire) increase with time, at least up to a certain limit—for example, within the season. Conversely, planning agents know that uncertainty of expectations at a certain future date will decrease as this date is approached in the course of time. We just concluded that one result of this relation between uncertainty and time is the desire to keep utilization plans flexible. Similarly, the discount for uncertainty increases with time. Because the dispersion around the most probable value of net revenues increases with time, net revenues with the same most probable value will require a greater and greater discount for uncertainty the further they are removed in the future.

Although uncertainty generally increases with time, there is no necessity that this increase is in the same proportion. Both introspection and observation of the behavior of planning agents would seem to indicate that uncertainty of expectations increases more than in proportion to time. In this connection, it may be well to emphasize again that uncertainty of expectations is a purely subjective force influencing the decisions of planning agents.

What about the relation of uncertainty preference to time? It would seem plausible to regard uncertainty preference as a character trait of planning agents which is not related to time. As a first approximation, uncertainty preference of an individual may be regarded as constant over time.

Allowance for uncertainty is influenced both by uncertainty of expectations and uncertainty preference. Thus, allowance for uncertainty will also increase with the distance expectations are removed in the future, and, probably, this increase is greater than in proportion to distance.

Uncertainty in Relation to Conservation Decisions

On the basis of the relation between uncertainty and time, we can now approach the effects of uncertainty as an economic force influencing conservation decisions. In other words, we may ask: How does a change in uncertainty of expectations and in uncertainty preference affect conservation decisions? Are the results conservation or depletion? Why do these results occur and what is their significance for conservation policy?

In discussing *changes* in the uncertainty of expectations we may assume that these changes apply to all intervals of the production plan after a certain date and are of a magnitude roughly in proportion to the previously existing uncertainties. Generally, this assumption will be quite realistic. Exceptions, however, may occur—for example, changes of uncertainty may be confined to one or a few intervals. In such cases no *general* statements about the effects on conservation and depletion can be made. One would have to know which particular intervals are affected and how these intervals are related to all other intervals through revenues and costs.

Under the assumptions made in the preceding paragraph, changes of uncertainty preference must have the same effects upon conservation decisions as changes of uncertainty of expectations. We will focus, therefore, upon the latter. Such changes may be allowed for, as we know, through uncertainty discount, hedging, pooling, spreading, and flexibility. It is best to discuss first the effects of changes of uncertainty discount and flexibility. Hedging, pooling, and spreading affect the utilization plan mainly by reducing the necessity of making uncertainty allowance through discounting and flexibility. Otherwise, hedging, pooling, and spreading are neutral with respect to the state of conservation.

Uncertainty Discount and Conservation

Because of the relations between uncertainty and time, changes in the uncertainty discount affect the state of conservation like changes of the time discount: an increase in the uncertainty discount means a progressive decrease (one that becomes greater with distance) in the present value of future net revenues. The result will be an attempt by planning agents to change the time distribution of net revenues in the direction of the present. Why such an attempt will lead to depletion was explained in chapter 7. Conversely, a decrease in the uncertainty discount will lead to conservation. The effects of changes in flexibility are not so simple. Before they are considered, another question must be answered.

The question arises whether planning agents in discounting for uncertainty use their own personal rates—which depend on their uncertainty preference *and* their estimates of dispersion—or whether such personal rates are coördinated through the market. This is the same question which had to be answered in connection with discounting for time.

The wide and varied field of insurance indicates that market coördination of personal uncertainty discount rates sometimes takes place. With a term taken from this field, we will call the market rate in discounting for uncertainty the "risk premium." Risk premiums stand in the same relation to personal uncertainty discount rates as interest rates to time-preference rates.

Relevance of risk premiums for individual utilization plans is dependent on the same conditions as were discussed in detail in the preceding chapter, namely, that markets for assets are perfect. This condition seriously limits the practical significance of risk premiums. Moreover, the actual situations in which risk premiums exist at all are more limited than situations in which interest rates exist. In the economics of conservation, risk premiums are even less significant in relation to personal uncertainty discount rates than interest rates are in relation to time-preference rates.

The uncertainty discount over a certain distance of time expressed in per cent and per unit of time [5] need not be the same as the time discount. A small individual time preference is frequently associated with a large uncertainty preference. According to our assumptions, the time discount increases as a first power of time. The uncertainty discount may increase as a higher power of time or as a more complicated function of time.

Flexibility and Conservation

Since planning agents are certain that uncertainty (dispersion) of expectations in future intervals will decrease as time goes on, they will attempt to avoid as far as possible those commitments which would make costly any future change in the utilization plan. A more flexible plan may yield smaller present net revenues than a fixed one, provided the most probable expectations underlying the fixed plan are realized, but it may yield greater present net revenues if these expectations are not realized. On the basis of the most probable expectations, therefore, flexibility involves decreases of revenues, or increases of costs, or both. If the costs of flexibility are high and if uncertainty is small, flexibility will not be carried far, and vice versa.

[5] The uncertainty discount is a function of uncertainty preference and of uncertainty (dispersion) of expectations. But since the latter is related to time and the former constant over time by assumption, the uncertainty discount may also be regarded as an increasing function of time.

Increased flexibility of utilization plans and increased uncertainty discount may be regarded as economic alternatives. However, these two alternatives are not equally effective in allowing for uncertainty—at least in many situations: First, the uncertainty discount cannot be used if the most probable outcome cannot be ascertained. Such situations are very common in economic reality. Second, the uncertainty discount may not be effective even if a most probable value can be ascertained. Such situations were mentioned in chapter 6; they exist if uncertainty involves the possibility of "immoderate losses" [6] (negative net revenues)—although the most probable net revenues may be positive and high. In such situations, discounting would not be an effective allowance for uncertainty if the contingency which creates losses should actually be realized. Greater flexibility, on the other hand, would be effective in guarding against such a contingency.

How is an increase in flexibility accomplished? Assuming a given command over assets of all kinds, flexibility is largely a problem of keeping liquid funds (cash, government securities) and liquid inventories (finished goods, raw materials) [7] and of reducing the time over which commitments in durable producer goods (land, buildings, equipment), in labor under some conditions (chapter 5), and in nonliquid inventories (semifinished goods) are made.

Planning agents can reduce such commitments by securing durable producer goods through short-term leases rather than through long-term leases and through ownership.[8] This, at least, is true as long as uncertainty about their tenure does not exercise a stronger influence than uncertainty about changes of technology, demand, and social institutions. The problem of insecurity of

---

[6] "Immoderate losses" we called (chap. 6) losses of a magnitude which threatens the continuity of a utilization plan.

[7] It may be noted that liquidity with the objective of allowing for uncertainty is not merely a question of holding cash (money). This type of liquidity would be ineffective in terms of the objective if, for example, the possibility of price inflation exists.

[8] Theoretically, they will also try to shift uncertainty bearing to the shoulders of creditors. However, there is little evidence that planning agents are successful in such attempts—except in securing government funds. Fixed charges are a deterrent in situations of high uncertainty and so are the costs of credit, which include an uncertainty discount by lenders.

tenure in relation to conservation will need special treatment. In any event, the tendency to shorten leases and to avoid ownership is limited by the costs of short-term leases in relation to the costs of long-term leases and of ownership. Planning agents, therefore, usually cannot avoid investment in *some* durable producer goods and nonliquid inventories.

If planning agents invest in such goods themselves, they can increase flexibility by giving preference to goods which are less durable. In the terminology used before, they will reduce the periods over which costs are sunk. A few examples may illustrate these possibilities of increasing flexibility.

A farmer who, as a result of competitive drilling and absence of a well-defined legal status of ground water, has become more uncertain about the dependability and costs of his future water supply, will hesitate to make new permanent improvements on his land and may even reduce maintenance and repair on existing ones; if he should make new improvements, he would receive no help from credit institutions.

A dairyman who wants to sell grade-A milk, but who has no sanitary barn and who is uncertain whether the comparative advantage of dairying over beef cattle or sheep will continue in view of labor conditions and government price regulations, will not erect a new barn, but will retain his old barn, which is adaptable for beef cattle and sheep, and will continue to produce process milk in spite of its lower price.

A forest owner who faces increased uncertainties in technology, demand, and government regulations will build less permanent roads and houses for his workers; he will also reduce replacement and maintenance of equipment—such as his sawmill.

Even expenditures of costs sunk for shorter periods of gestation —for example, in connection with fertilizing, terracing, improving of breeding stock, planting orchards and vineyards—will be reduced if there is an increase in uncertainty.

What does an increase in flexibility mean in terms of conservation? The answer depends on whether flexibility is accomplished by increasing liquidity or by reducing sunk costs.

An increase in liquidity does not in itself mean conservation or depletion—in the same way that disinvestment in resources in itself has no necessary relation to conservation and depletion. On the other hand, a reduction in the periods over which costs are sunk in durable producer goods and nonliquid inventories means depletion. Likewise, tendencies to avoid ownership and to shorten leases generally result in depletion—as will be discussed in more detail later.

Usually—although not necessarily (namely, if liquidity alone is the method of adaptation)—changes of uncertainty, if taken into account through flexibility, tend to change utilization plans in the same direction as if such changes are taken into account through discounting for uncertainty. In other words, with some exceptions (see also pp. 126–127), an increase of uncertainty tends toward depletion and a decrease toward conservation.

Hedging and Conservation

Hedging means the shifting of uncertainty bearing to professional specialists, usually to speculators in a futures market. These specialists may have greater knowledge in taking uncertainty into account through proper statistics and estimates, or a smaller uncertainty preference, or both.

Existing futures markets do not extend far enough into the future to make hedging important in the economics of conservation. Most futures markets extend not more than a year into the future. They are important for traders in resources already produced, but not for producers in their conservation decisions. There is little possibility that this situation will ever change: Producers of resources have to sink costs before the resulting revenues come in. They are anxious, therefore, to hedge until they recapture these costs through revenues. They expect to recapture a larger and larger portion as time goes on. For more distant intervals, therefore, their incentive to hedge, that is, the supply of futures, will decrease.[9] Buyers of futures, on the other side, have no similar previous commitments. They will buy futures only at a price sufficiently lower than their own expectations to permit a specula-

[9] To some extent, an increase in flexibility (through recovery of sunk costs) and hedging may be economic alternatives in making allowance for uncertainty.

tive profit. The more remote the interval for which contracts are made, the greater is the uncertainty of buyers, and the lower the price they are willing to contract for immediately. The demand for futures therefore decreases the more, the farther intervals are removed in the future. Because of this situation in the supply and demand for futures, the present price for more distant future deliveries soon becomes so low that producers will prefer not to hedge. Before this point is reached, numbers of buyers and sellers become too small for a perfect market.

Within the narrow limits of its effectiveness, hedging reduces the need for uncertainty allowance through discounting and flexibility. Hedging, therefore, encourages conservation. For example, hedging would make it safer—as compared with nonhedging—to sink costs for fertilizer and better cultivation practices in crop production. On the other hand, speculative expansion of crop production, that is, expansion dependent on the expectation of an especially favorable price situation—excluded by hedging—would be discouraged. A speculative expansion of production usually consists of an expansion of acreage, with a minimum of sunk costs per acre, into areas which under more conservative price expectations would not be suited climatically or topographically. Such speculative expansion of crop acreage has frequently led to serious soil depletion.

Pooling and Conservation

Through pooling, some uncertainties which are important for conservation can be more effectively reduced than through hedging. With pooling, in contrast to spreading discussed below, uncertainties are of the same kind but of random incidence among producers in the same statistical class—that is, operating under similar physical and economic conditions.

The most important condition for pooling is that the membership in the pool be sufficiently large. Since the precision of an average is proportional to the square root of the number of terms it contains, pooling enables the most probable value of expectations to be determined more precisely (that is, with smaller dispersion) than is possible for individual planning agents. As a corollary, the aggregate allowance for uncertainty in the pool—that is, the sum of risk premiums charged to pool members—is smaller than the aggregate of individual uncertainty allowances without pooling.

Some uncertainties—for example, those connected with fire, hailstorms, and pests—conform to the conditions of pooling. Pooling arrangements are actually made in many of these cases. Expansion of such arrangements appears desirable, not only from the standpoint of reducing the aggregate of uncertainty allowance in the community, but also from the standpoint of resource conservation, especially in agriculture and forestry. Public policy can facilitate such expansion by clarifying the statistical problems involved and by enacting regulations to insure that pools are sufficiently large. In some countries (Japan, U.S.S.R., Canada) crop insurance has been made compulsory. In the United States, the Federal Crop Insurance Act passed in 1938 is an example of a voluntary government-assisted and government-subsidized insurance scheme.

Some uncertainties are connected with contingencies which are random only over a considerable period of time. Such uncertainties are, for example, drought, frost, and floods. In these cases pooling may reduce the aggregate allowance for uncertainty provided the pool is continuous over time. The pool will be better able than individual planning agents to secure an expert staff for analyzing the statistical problems involved and to guard against contingencies by accumulating reserves and securing credit. In this case of pooling, government assistance is needed even more than in cases where uncertainties are random over shorter periods of time (fire, hailstorms, pests). The statistical problems are more complex and the need for reserves and credit much greater.

An example for this situation is drought in specific regions, let us say, in the Great Plains area. Drought is random over time only if presently available series of data are considerably extended. Even weather data—which are more accurate and extend over longer periods than yield data—do not as yet establish a sufficient actuarial basis for dealing with a major drought. Experience shows that drought years tend to bunch. A period over which this bunching can be called random is considerably longer than presently available series.

What about the relations between pooling and conservation? If contracts (insurance policies) between the pool and individual resource users or, at least, with resource users in the same statistical class, were entered into on a strictly actuarial basis, we could draw conclusions similar to those for hedging: the need for uncertainty allowance through discounting and flexibility would be reduced, expenditure of sunk costs encouraged, and speculative expansion of production discouraged; in other words, pooling would tend toward conservation. However, the difficulties of making pooling arrangements on a strictly actuarial basis are great, as was mentioned previously. As a matter of fact, contracts between individual resource users and the pool are frequently made not on an actuarial basis but are more favorable to the individual resource user than is warranted on such a basis. The loss to the pool is made up by government subsidies.

Government subsidies to pooling may actually encourage speculative expansion of crop acreage into areas with higher production hazards connected with climate, soil, or topography. Furthermore, government subsidies introduce an element of uncertainty: producers are usually uncertain about the continuity of government subsidies over time in view of changing political and economic conditions. They will be tempted to expand speculative production in the immediate future, for which they are fairly certain of government subsidies, even if such an expansion means resource depletion. Thus, a subsidized pooling arrangement may actually be an influence toward depletion.

Spreading and Conservation

Some uncertainties are connected with unique (nonrecurrent) contingencies which affect all producers in the same statistical class, but those in different classes differently. Such uncertainties are connected, for example, with changes of technology and demand and with the discovery of new deposits in the case of stock resources. These uncertainties cannot be reduced through pooling by producers in the same class. Often, however, these uncertainties can be reduced for individual planning agents by spreading.

Spreading is possible if different branches of a multiple-product firm, different firms, different industries, or different geographic areas, are subject to uncertainties of different kinds. Aggregate uncertainty allowance for such a composite of uncertainties can be less than the aggregate allowance for individual kinds of uncertainties because the dispersion of possible around the most probable (and actual) revenues and costs is less for the composite than for individual components. It is the principle of not putting all eggs into the same basket. The spreading of uncertainties within a multiple-product firm is especially important in diversified agriculture. Spreading of uncertainties within an industry, within a country, and even internationally, is common in diversified insurance companies and banks.

Spreading, like hedging and pooling on an actuarial basis, reduces the need for uncertainty discount and flexibility. If insurance premiums to different components are properly adjusted in relation to their contributions toward the composite uncertainty, speculation is discouraged. Thus, spreading favors a more conservative production plan.

On the other hand, if producers do not have to pay a proper premium, if it is paid largely by the government, spreading of uncertainties within the enterprise will be discouraged, speculative expansion of production encouraged, and a new element of uncertainty, the one just described for subsidized pooling arrangements, introduced. Under these conditions, insurance based on spreading may favor resource depletion.

Shifting of Uncertainty Bearing to the Public

Uncertainty bearing is not necessarily shifted to the public merely because the public assists the hedging, pooling, and spreading of uncertainties, through regulations, or even directly undertakes this task on the basis of actuarial principles.

If such a shift takes place, it may be socially desirable as a form of subsidy to certain groups, especially in emergencies; but unlike the mere participation of the public in hedging, pooling, and spreading, it has a debit side that has already been mentioned: Public credit and confidence may be impaired. In consequence, uncertainty may be increased in private utilization plans.[10] Diversification within firms may not take place, although this may be the most economical way to bear uncertainty. This may happen, for example, in a crop-insurance plan for monocultures (wheat, cotton) in which farmers do not pay the full premium. More serious maldirection of productive services may result if uncertainty bearing by the public takes the form of guaranteed prices and incomes.

From the standpoint of conservation, as distinct from income redistribution, these policies are of value if they result in a reduction of uncertainty of private planning agents. In the past, they have been more successful in changing the income distribution.

Uncertainty and Private Capitalism

On one hand, resource utilization under private capitalism is subject to greater uncertainties than under a social organization that is, to use Sombart's terms,[11] precapitalistic or acapitalistic. Hazards of nature and changes in number of consumers and in technology operate under any social organization. But uncertainties caused by inflationary and deflationary changes of consumers' incomes, by changes of fashions, and by certain erratic actions of producers characteristic for monopolistic markets, are especially important under capitalism. The tendency toward depletion that capitalistic resource utilization frequently exhibits, if compared with other forms of social organization, is caused, partly at least, by greater uncertainty and by a calculative allowance for this uncertainty—instead of relying on habit patterns.

---

[10] It scarcely needs to be emphasized that the kind of government expected is highly important for planning agents. Instability of government at all levels (national, state, and local) and vagueness, indecision, yielding to pressure groups, and contradictions in policies cannot fail to increase uncertainty in private utilization plans.

[11] Werner Sombart, *Der moderne Kapitalismus* (München und Leipzig: Duncker und Humblot, 1928), 3 vols.

On the other hand, in recent decades, there is a trend observable in nearly all capitalistic countries, relatively late but strong in the United States, to decrease uncertainty of private planning agents through better statistics and estimates (agricultural extension, outlook reports), through tenure legislation, credit policy (loan guarantees, ceiling on interest rates to resource users), price policy (minimum prices, parity prices, crop loans, guaranteed volumes of sale), many forms of insurance schemes (crop, livestock, fire), and other national and international policies or proposals (buffer stocks, commodity reserves, and currencies) that may reverse tendencies towards greater uncertainty and, therefore, depletion under capitalism as compared with a precapitalistic or acapitalistic social organization.

If private capitalism is compared with a social organization in which economic activity is controlled by the authority of the modern state—rather than with precapitalistic or acapitalistic social organizations—some important new factors will increase uncertainty allowance.

By selection, training, and experience, the appointees of a centralized bureaucracy will have a greater (negative) uncertainty preference than private planning agents. Those sectors of the economy in which private economic decisions are permitted to function are more strongly affected by the uncertainties regarding the government's actions. These political factors are strengthened if the state authority is a dictatorship: periodic "purges" of the bureaucrats will increase their fear of making commitments. Governmental action tends to be more extreme in its consequences when democratic checks and balances are inoperative. Thus, as in many other fields, state authority does not offer a sure and simple short cut for decreasing uncertainty allowance in economic decisions.

Uncertainty and the End of the Planning Period

It may be noted that uncertainty, like interest, is neither a necessary nor a sufficient condition for bringing the planning period to an end. True, allowance for uncertainty, like interest, weakens progressively (that is, increasing with distance) the influence of future net revenues upon shape and extent of the production plan; uncertainty helps to bring the planning period to an end sooner than otherwise would be the case. But usually, the planning period would come to an end (both in terms of intervals and of clock time) even if uncertainty did not exist.

It is possible also to imagine a production plan leading to a stationary state in spite of uncertainty. In this case, the extent of the planning period would not be a finite magnitude—at least in terms of clock time.

However, if allowance for uncertainty is made through discounting and flexibility, the influence of future net revenues upon shape and extent of the production plan becomes small beyond a certain distance. It is not worth while to spend time and effort in ascertaining net revenues in additional intervals. In practice, therefore, a somewhat arbitrary maximum number of intervals can frequently be chosen although the theoretically optimum number of intervals may be larger. It was suggested earlier that the influence of uncertainty upon production plans increases with time more than that of interest. The maximum number of intervals chosen because of uncertainty is therefore usually smaller than that chosen because of interest. In the examples mentioned in chapter 6, the maximum number of intervals chosen because of interest was around fifty (years). In private enterprises with high uncertainty, the maximum may be ten (years) or even less.

The effects of uncertainty upon the production plan need a qualification which has some bearing on the relevance of a finite planning period. This qualification is connected with habit patterns.

Uncertainty and Habit Patterns

If uncertainty is great in relation to the knowledge of allowing for it on the basis of statistics, estimates, and hunches, the calculative approach to the optimum state of conservation may be abandoned. Planning agents may fall back on custom, tradition, and following the leader: in short, they fall back on what we called "habit patterns."

For example, a farmer may not conceive of the variability of weather as a contingency which can be made more definite through proper statistics; or he may not conceive of shifts in technology, demand, and institutions as hazards about which helpful (in planning) estimates can be made, but as acts of God which just have to be taken in stride. If he has this attitude, he may not take account of uncertainty through greater flexibility of utilization plans but, on the contrary, through greater rigidity, that is, by following habit patterns.[12]

The tendency to take refuge in habit patterns, when uncertainty is great in relation to the possibilities of allowing for uncertainty, is accentuated by another factor: There is some historical indication that periods of great political and economic uncertainty are favorable to the growth of social institutions that take a large share of responsibility for independent decisions away from individual planning agents in favor of a centralized bureaucracy, whether government, church, or guild. Such institutional arrangements tend to rely on and to strengthen habit patterns.

Habit Patterns and Conservation

As already stated, habit patterns may favor either conservation or depletion. Which they favor depends on the particular habit pattern and on the particular physical and economic environment that is being studied.

In societies that have existed in the same environment over long periods of time, habit patterns have frequently been "selected" that safeguard a particular state of conservation—a state that is of great interest when studying the influence of uncertainty in the social economics of conservation. This particular state we may call a "safe minimum standard" of conservation. It is relevant for that important class of flow resources the utilization of which is characterized by a critical zone (chapter 3). In chapter 18 we will deal in greater detail with this standard.

In farming areas occupied by a stable homogeneous peasant population with an active and closely knit cultural life, standards

[12] Allowance for uncertainty through discounting, hedging, and insurance schemes is ruled out by our assumption. However, habit patterns may incorporate primitive pooling and spreading arrangements.

of good husbandry (at or above the safe minimum) in the field of soil conservation become well defined and accepted. This is true for peasant farmers of western and central Europe, for religious groups among the Pennsylvania Dutch, and for the Japanese rice grower.

Habit patterns, however, do not always safeguard a minimum standard of conservation. Sometimes habit patterns establish a state of conservation which is below the safe minimum standard. This happens, for example, if people migrate (voluntarily or by force) into new areas where climate, topography, or soils are not suited for practices based on the old habit patterns. The westward migration in the United States affords many interesting examples of depletion caused by the institutional baggage of the migrants.[13]

[13] S. V. Ciriacy-Wantrup, "Land Use, the Basis of Western Economy," *The Westward Migration and Its Consequences* (Divisional Symposium. Annual Meeting, Pacific Division, American Association for the Advancement of Science, Salt Lake City, 1950), 12 pp. Processed.

CHAPTER 9

PRICES AND PRICE SUPPORTS

Difficulties in Economic Analysis

The preceding two chapters tried to show how changes of two particular market prices—interest rates and risk premiums (and changes of the corresponding individual preferences)—tend to affect conservation decisions. These changes have an important functional relation to time: The effects of any given change increase with time over all affected planning intervals. The effects of any given change upon interrelations of use rates (through marginal revenues and costs) are in a definite relation to time—for example, periods over which costs are sunk are either shortened or lengthened. Because of these relations to time, we could deduce fairly clear-cut effects upon conservation. For this reason, interest and uncertainty were discussed first.

When we now turn to the effects of changes of expected prices of products and of productive services in general, difficulties in economic analysis and exposition become formidable.

First, a given change of prices may affect equally all the planning intervals over which it lasts; its effects do not increase with time, as do the effects of any given change of interest and uncertainty. A price change that affects equally all intervals of the planning period would give no incentive to change the time distribution of use rates (provided interrelations of use rates through marginal revenues and costs are not affected). However, it is rare that a price change is expected to last over the whole planning period. If price changes do not affect all intervals equally, what are the effects of different time distributions of price changes?

Second, interrelations of use rates through marginal revenues and marginal costs may be affected in a great variety of ways depending on what productive services and what products are affected. For example, it makes a difference for soil conservation whether the price of fertilizer or the price of plows or disks changes. It is not immaterial for the conservation of forests whether the prices of equipment needed in logging with donkey engine and railroad change in relation to the prices of logging trucks and tractors. In multiple-product enterprises like agriculture, conservation decisions may be affected differently if the price change is in tobacco or alfalfa, cotton or meat, milk or wheat.

Thus, in order to make conclusive statements about the effects of price changes upon conservation decisions, we would have to know, first, how these changes are distributed over time, and second, how interrelations between use rates in different intervals through (marginal) revenues and costs are affected (chapter 5).

Simplifying Assumptions

The complexity created by these two requirements can be reduced by making simplifying assumptions with respect to the time distribution of price changes and with respect to the effects of these changes upon complementarity, competitiveness, and independence of differently dated use rates.

For example, with respect to time distribution one may assume that a change of prices is expected to occur at a certain future interval and to last indefinitely thereafter (or, at least, for the whole planning period); or that prices are expected to change over time in the same direction at a constant or increasing rate.

With respect to interrelations of use rates through revenues, it was mentioned in chapter 5 that in the utilization of flow resources (agriculture, grazing, forestry) independence in revenues can generally be assumed for all intervals.

With respect to interrelations through costs, one may assume that price changes either strengthen or weaken complementarity or competitiveness through costs over the whole extent of the production plan.

Many other simplifying assumptions can be made. For example, it may be assumed that price changes take place in certain intervals only, or that such changes have a certain (cyclical) sequence.

Let us observe then how price changes affect conservation decisions under simplifying assumptions. We cannot cover all *possible* assumptions. We can, however, explore the effects of price changes upon conservation decisions under assumptions which are of special practical significance. (It need scarcely be mentioned that we are considering *relative* price changes—that is, prices other than the ones discussed are held constant.)

Effects of Different Time Distributions of Price Changes

Let us first observe the effects of different time distributions of price changes under the assumption that interrelations of use rates through revenues and costs are not affected.

An increase of product prices which is expected to occur at some future interval and to last indefinitely, or an increase of product prices which is expected to become greater with time, will induce planning agents to shift rates of use toward the future—that is, will result in conservation. An expected decrease of product prices under corresponding assumptions will lead to depletion. There is some indication that in economic reality a change of prices often leads planning agents to expect further changes in the same direction in the future; in the language of the economist, the elasticity of their price expectations is frequently greater than unity.[1] For this reason, the above assumptions are not of theoretical interest only.

---

[1] The elasticity of price expectations may be formulated by:

$$\frac{\text{proportional change in future price}}{\text{proportional change in current price.}}$$

On the other hand, under other assumptions of equal practical significance, quite different conservation effects will result. For example, if a current increase of product prices is not expected to last longer than a given period—let us say three or four years (as during a war or during the upswing of a business cycle)—planning agents will try to shift use rates toward that period; if longer periods are planned for, or are considered longer by the observer, such a shift means depletion—*ex ante* or *ex post.*

Sometimes, the effects of price changes appear to contradict our conclusions: an increase of product prices—even if it is not limited over time in the expectations of planning agents—may result in depletion of rented agricultural or grazing lands. This was the experience on public grazing lands before effective controls were available. As we shall see in chapters 10 and 11, it is analytically more desirable to regard such depletion as not due to price increases *per se* but to the effects of indefinite tenure.

With price changes of productive services, the effects of different time distributions tend in the opposite direction from that just observed for products. However, in this case, the effects on interrelations of use rates through costs are usually so great that attention must be focused on these effects rather than on those created by different time distributions of price changes.

Price Changes of "Conserving" and "Depleting" Services

Planning agents employ most productive services in conservation as well as depletion. In agriculture, for example, labor and equipment (tractor, plow) may be used for exposing slopes to water erosion or semiarid grasslands to wind erosion; the same labor and equipment may be used for building terraces or for contour furrowing. Powerful earth-moving machinery (bulldozers, shovels, scrapers) may be used for stopping gullies or for clearing land and for strip mining.

Thus, in order to state conclusively how price changes of any one of these productive services affect conservation decisions, we need assumptions with respect to the technological conditions and practices under which it is employed.

For some productive services we may assume that they are conserving under most technological conditions and practices. In agriculture, for example, fertilizer, purchased feeds, legume and grass seeds, and the crawling tractor with its various attachments are usually (but not necessarily) of this type. Some other services are usually (but not necessarily) depleting; examples are machinery employed in mining,[2] the donkey engine in forestry, and most hunting equipment.

In principle (and according to our definitions of conservation and depletion), only the production plan and a practice are "conserving" or "depleting." However, use of these terms in connection with some productive services is a convenient shorthand expression. Such use implies that conserving services always strengthen complementarity (or weaken competitiveness) in costs between use rates. The opposite is implied for depleting services. With this precaution in mind, we may say that increases in the prices of conserving services or decreases in the prices of depleting services will lead to depletion; decreases in the prices of conserving services or increases in the prices of depleting services will lead to conservation.

Price Changes of "Conserving" and "Depleting" Products

Frequently, in conservation literature, the terms "conserving" and "depleting" do not refer to productive services but to products. This terminology has become particularly important in the economics and policies of soil conservation. Here, likewise, caution is advisable in interpreting the terms "conserving" and "depleting."

In diversified farming (or other multiple-product enterprises), there are crops (products) which usually strengthen complementarity in costs (or weaken competitiveness); there are other crops which usually weaken complementarity in costs (or strengthen competitiveness). Examples of the former type of crops are alfalfa and grasses; examples of the latter, tobacco and cotton. Alfalfa and grass may be called "conserving" and tobacco

[2] Under some technological conditions this is not true. Mining machinery, for example, that makes it economical to follow thinner veins or to leave less ore in support of the overhang would be conserving as compared with some other equipment.

"depleting." A price increase of these crops and, consequently, a relative expansion of their acreage *may* result in soil conservation and depletion respectively.

In these cases the terms "complementary" and "competitive" refer to relations between different crops (products) *and*—derived from these relations—to the relations between different intervals. In measuring the latter relations, production, costs, and revenues of the whole crop rotation must be considered.[3]

"Complementarity" and "competitiveness" in this sense change if one of the crops in the combination is changed: As stated above, an expansion of grasslands at the expense of cultivated acreage usually leads to soil conservation. But an expansion of grasslands at the expense of forests may lead to soil depletion. Relative expansion of tobacco and cotton acreage usually leads to soil depletion. However, expansion of tobacco and cotton at the expense of small grain may be beneficial for soil conservation provided certain conservation practices (contour cultivation, terracing, fertilizing) are used and expand with the tobacco and cotton acreage.

For many crops other than those mentioned in the preceding paragraph, the terms "conserving" and "depleting" are even less adequate. Sugar beets and potatoes, for example, may be conserving in one system of land use, depleting in another. An increase of the prices for sugar beets and potatoes would bring about soil conservation in European farming; [4] in some parts of the United States a similar change of prices may lead to soil depletion. The same is true for some legumes such as peas and soybeans. In many parts of the world these two crops are highly important in soil conservation; on the other hand, the increase of the prices for peas during the First World War was a big factor for soil erosion in the Coast Range of California, where pea acreage expanded at the expense of permanent grasslands; and some soil experts

[3] Some meaningful common denominator for production (rates of use) must be found in this case: quantities of different crops must be weighted by calories, prices, and the like.

[4] The reasons were explained in S. V. Ciriacy-Wantrup, "Soil Conservation in European Farm Management," *Journal of Farm Economics*, Vol. 20, No. 1, February 1938, pp. 86–101. (University of California, College of Agriculture. Giannini Foundation of Agricultural Economics, Paper 66.)

feel that the expansion of soybeans under the stimulus of prices during the Second World War went too far from the standpoint of soil conservation under the farming systems prevailing in the Middle West.

In these and similar cases of soil depletion in the United States, inclusion of a particular crop in a production plan or an increase in its acreage was neither a necessary nor a sufficient condition for soil depletion. To subsidize prices of "conserving" crops and to impose penalties for growing "depleting" crops is, therefore, not always in the interests of conservation. On the contrary, the general classification of individual crops (rather than of utilization plans, practices, and farming systems) as "conserving" and "depleting" confuses theoretical and practical issues and may lead to the waste of public funds. An example taken from the United States soil-conservation policy may illustrate these points.

Penalties for "Depleting" Products

Before the last war, certain crops were stigmatized as "soil depleting," and penalties (as deductions from AAA-benefit payments) were imposed for their cultivation. Until 1941, corn grown for silage, sugar beets, grain sorghums, peanuts, potatoes, truck and vegetable crops, field beans, peas, soybeans harvested for seed, flax, and summer fallow, among others, were officially classified as "soil depleting." [5]

Some of these crops are often valuable for soil conservation because they may serve as a basis for a more diversified livestock economy, or because they may be helpful in eradicating weeds through row cultivation and shading effects, or because they can be fertilized heavily without danger of lodging. With the growing demand for crops during the Second World War, the label "depleting" was generally taken off these crops. The conclusion seems warranted that other objectives than conservation—for example, maintenance of parity prices—were responsible for the identification of conservation with a reduction in the acreage of these crops.

---

[5] See U. S. Department of Agriculture, U. S. Agricultural Adjustment Administration, Office of the Administrator, *1941 Agricultural Conservation Program Bulletin* (Washington: Govt. Print. Off., 1941), pp. 33–35.

Price Supports and Conservation Policy

When considering the relation between income and conservation decisions, we concluded that support of resource user's income, especially in the low-income groups, is a more effective form of public assistance—at least from the standpoint of resource conservation—than support of product prices. This conclusion is strengthened by the relations discussed in this chapter: Without definite assumptions with respect to time distribution of supports and their effects upon interrelation between use rates, such supports are an uncertain tool of conservation policy. Moreover, supports are more likely to bring about depletion than conservation. This last point may be elaborated somewhat further.

Price supports are not likely to change expectations of resource users in the fashion assumed above; usually, they are not expected to last indefinitely or to increase with time. Price supports are generally dependent on the political fortunes of a government, a party, or a pressure group. Price supports, therefore, will generally terminate (in the expectations of resource users and in reality) at a certain date—for example, on the date another party takes the reins of government—or, at least, receive a heavy allowance for uncertainty for all intervals after that date. Price supports, therefore, will induce resource users to shift use rates toward those intervals in which price supports are more certain—that is, toward the present. This is depletion. The effectiveness of price supports as a tool of a depletion policy may be increased through a public statement by the government that supports will be discontinued at a certain date.

Under some conditions—for example, in order to speed up the utilization of stock resources during a war—depletion is a policy objective. On the other hand, as a tool of a conservation policy—for example, in agriculture—price supports are suited only in those special cases where prices of a bona fide conserving service or product are supported.

Price Supports versus Income Supports

Income supports are no different from price supports with respect to "political" uncertainty. However, income supports are not directly related to rates of use. They can easily be given in a way that is neutral with respect to conservation decisions or, if desired, in such a way as to aid in conservation.

This relative superiority of income supports from one particular viewpoint, however, does not mean that they are superior from other viewpoints. For example, it may be argued that income supports, because of difficulties in determining their base and their changes over time, are more subject to political pressure than are price supports which operate automatically once the support in terms of market price is agreed upon. The politics of income supports may do more damage to the roots of a free economy than those of price supports.

Furthermore, the relative superiority of income supports from a conservation viewpoint does not mean that there are no better alternatives for raising incomes than government supports to either price or income. Neither price supports nor income supports go to the cause of low incomes. Do identical productive services earn less in one type of economic activity than in others? If they do not earn less, is their combination (size of the enterprise) too small to earn an income comparable with combinations in other fields? If differentials in income productivity exist, what is their relation to time? What economic and institutional forces are responsible for such differentials? In what way do such forces change the opportunity of planning agents to overcome income differentials? Are there possibilities of modifying the effects of these forces through tools of public policy?

These questions will not even be raised, much less answered, if reliance is placed on government supports—either to income or to prices. In the following chapters an attempt will be made to answer some of these questions.

Effects of Price Changes: Conclusion

The absence of hard and fast rules about the effects of price changes upon conservation decisions may cause disappointment among some readers. However, emphasis on this absence and on the importance of specific assumptions with respect to the time distribution of price changes and their effects on interrelations between use rates seemed necessary, because these assumptions are so frequently overlooked.

Price changes of products and of productive services have far-reaching effects upon conservation decisions; but to ascertain the direction of these effects (that is, conservation or depletion) and to assess their quantitive significance requires great caution. Sometimes, the concepts of conserving or depleting products and services are helpful as a first approximation in understanding and predicting results. However, care should be taken to make the assumptions which are implicit in such a terminology explicit in terms of technological conditions and practices. This is especially true if price changes are advocated as a tool of conservation policy.

CHAPTER 10

PROPERTY

Economic Forces and Social Institutions

The economic forces discussed in the preceding three chapters are strongly conditioned, and in some cases one might say generated, by social institutions. In this and the following chapters the most important of these institutions will be discussed from the standpoint of their influence upon private conservation decisions. Such a discussion is more than merely a different focus on the economic forces discussed so far. Social institutions affect conservation decisions in many ways other than through interest, uncertainty, and prices. An example may illustrate this point.

The social institutions determining property rights in resources may have important effects upon conservation decisions through uncertainty, if property rights are unstable over time. Still, there are many other ways in which property rights affect conservation decisions: the incidence of revenues and cost between owner, user, creditor, worker, and the public is largely determined by property rights, and influences conservation decisions greatly.

There is another reason why social institutions require particular attention in the field of economic forces affecting conservation decisions: social institutions can be changed. Social institutions, therefore, are of fundamental interest for conservation policy. For conservation policy, they may be tools as well as obstacles.

A more searching inquiry of how social institutions are related to public policy will be undertaken later. However, at this point differentiation must be made between two basic aspects of social institutions—their "concept" and their "structure."

Concept and Structure of Social Institutions

W. G. Sumner has presented a classic statement of two basic aspects or elements of a social institution: "An institution consists of a concept (idea, notion, doctrine, interest) and a structure. The structure is a framework, or apparatus, or perhaps only a number of functionaries set to coöperate in prescribed ways at a certain conjuncture. The structure holds the concept and furnishes the instrumentalities for bringing it into the world of facts and action in a way to serve the interests of men in society." [1]

In this part of our study we are mainly interested in the structure of social institutions—for example, in the laws, ordinances, regulations, administrative and judicial processes, customs, and mores which bring property rights, as actual economic forces, into the daily lives of individual planning agents. We cannot avoid touching upon the concept of social institutions in dealing with private economics of conservation. For example, a social purpose of property rights must be assumed. However, a more detailed discussion of the concept fits better into the analysis of social economics of conservation in Part IV.

Property as a Primary Economic Institution

Social institutions are important in many fields of human relations—for example, in the fields of matrimonial and domestic relations, of religious worship, of international political intercourse, and of economic behavior. Although conservation decisions are affected by social institutions in numerous fields, attention must be focused on institutions that affect economic be-

---

[1] William G. Sumner, *Folkways; a Study of the Sociological Importance of Usages, Manners, Customs, Mores, and Morals* (Ginn and Co., 1906), pp. 53–54.

Cf. also William G. Sumner and Albert G. Keller, *The Science of Society* (New Haven: Yale University Press, 1927 ; London: Oxford University Press, 1927), 4 vols.

havior directly or, for short, on economic institutions. Limitation of space makes such a restriction unavoidable.

Among economic institutions, property rights in resources are of paramont importance in the economics of conservation. Property rights can be called a primary economic institution not only because of their importance but also because several kinds of economic institutions are "derived" from them in the sense that their concepts and structures are variations of a common theme. Such derived property institutions are mainly the systems of tenancy, credit, and taxation.

Property has been aptly called a "bundle" of rights to control. The individual "strands" of this bundle may be distributed among the organized public (the state), the owners, the users, the creditors, the laborers, and possibly others. In this study, the term "tenure" will be applied to *all* strands of the bundle of property rights.[2] It will be used for all relations of control between men and resources. The term includes, therefore, tenancy, credit, and taxation (and other forms of public control). The term "tenancy," on the other hand, deals with those strands of the bundle of property rights which are surrendered by the owner of resources (who may be a public agency) to the user.

The specific influence upon conservation of the three "derived" property institutions warrants individual consideration (chapters 11 to 13). In this chapter we will deal in general terms with the broad over-all relations between property rights and conservation. These relations will be discussed under "indefiniteness," "instability," and "imbalance" of property rights.

Indefiniteness of Property Rights

Indefiniteness of property rights exists in that important class of resources which the user has to "reduce to possession"—in legal terms—before he owns them. These resources are usually called "fugitive" because they must be "captured" through use. Important fugitive resources are wildlife in the United

[2] "Tenure" meant originally the relation between grantor and grantee of land in the feudal system. The term is still used largely in connection with land, and there is no agreement in economic and legal literature about its comprehensiveness.

States, migratory waterfowl, high-sea fisheries, range forage on the public domain before the Taylor Grazing Act, and oil, natural gas, and ground water under tenure conditions in which overlying surface lands are in several separate ownerships and where control of subsurface resources is vested in the surface owners.

Resource tenure is not well defined in these cases. Definite property rights belong only to those who are in possession—that is, who get there "fustest with the mostest." Every user tries to protect himself against others by acquiring ownership through capture in the fastest possible way. Deferred use is always subject to a great uncertainty: others may capture the resource in the meantime. If the allowance for this uncertainty of tenure is great, concentration of all use in the smallest possible number of planning intervals near the present will become the most economical utilization plan for the individual planning agent. Under more definite tenure such depletion would appear highly uneconomical. For an individual trying to capture fugitive resources in competition with others, economic calculations become similar to those relevant for stock resources with rapid natural deterioration.

The wasteful depletion resulting from this state of affairs is common knowledge as far as wildlife and grazing resources are concerned. In these cases public opinion is now backing adequate remedial measures. The continuing wasteful depletion of underground fugitive resources has made a much smaller popular impression. The technological aspects of remedial measures are more complex. However, such problems as the proper spacing of wells, the proper gas–oil ratio in pumping, the great possibilities of increasing extraction rates by injecting water or gas under high pressure into oil pools,[3] the draft-replenishment balance of an aquifer, and the increase of replenishment through spreading are well known to the specialist.

[3] At present, average extraction rates are not more than 45 per cent. In some pools under unregulated competitive drilling, for example, Placerita Canyon in southern California, extraction rates may be as low as 25 per cent. Through pressurizing oil pools these rates could be boosted to 90 per cent. However, the process is costly, and every competitor using the pool would profit from it.

Remedies for Indefiniteness of Property Rights

There are two main remedies to the wasteful depletion caused by indefiniteness of tenure. First, control over resource use may be defined through law and government regulations in such a way that the need for capture disappears. Second, instead of making private tenure more definite, control of resource use may be vested in the government.

The first remedy is used with respect to wildlife in some European countries and with respect to oil, natural gas, and the public range in the United States.

In some European countries, wildlife is regarded as belonging to the owner of the land surface on which the animal happens to be. The owner may then rent out his use rights for a compensation.[4]

In the United States, some attempts have been made to vest the ownership of oil and natural gas collectively in all those who have drilling rights in a certain oil pool and to have the collective owners establish unified control over the utilization of the pool. Examples of unitized pools are Kettleman Hills and Domingues in California and the great pool in east Texas.[5] In such pools the

---

[4] In the United States the right to wildlife can be secured by owners of surface lands only indirectly and imperfectly through the trespassing laws. Recently a development similar to that in Europe got under way in the United States—the fast-growing importance of game-management areas. See U. S. Congress, House Select Committee on Conservation of Wildlife Resources, *Wildlife Conservation: Report No. 1 of the Select Committee on Conservation of Wildlife Resources* (Washington: Govt. Print. Off., 1935), 36 pp. (74th Cong., 1st sess.).

[5] Other pools in California are not unitized. A recent spectacular example of the effects of indefinite property rights is the Placerita Canyon in southern California, mentioned above (footnote 3). As a result of competitive drilling, only about 25 per cent of the estimated 60 million barrels of oil in the Placerita field will be recovered; 10 billion cubic feet of natural gas has already been lost.

In California unified control is voluntary. The California state oil-conservation law providing for spacing of wells was declared unconstitutional. Unified control means production allotments to each well in combination with proper spacing of wells. Production allotments consider such factors as gas-oil ratio, depth of the well, and use of the lot on which the well is drilled. See:

U. S. Federal Oil Conservation Board, *Report V of the Federal Oil Conservation Board to the President of the United States* (Washington: Govt. Print. Off., 1932), 60 pp.

Leonard Logan, *Stabilization of the Petroleum Industry* (Norman: University of Oklahoma Press, 1930), 248 pp. (Oklahoma Geological Survey, Bull. No. 54).

Rule of Capture [6] is replaced by the Correlative Rights Doctrine, developed in California for adjusting conflicting claims to percolating ground water.[7]

A similar unified control was established over the public range after the Taylor Grazing Act of 1934 had provided the legal and administrative basis.[8] The Act ended the fugitive nature of the Western range resources and terminated a period of drastic depletion. Conservation was accomplished by (1) the withdrawal of public domain, (2) the establishment of grazing districts, (3) the administering of such districts by the Grazing Service of the Department of the Interior in coöperation with local stockmen, (4) the issue of grazing permits to individual users. Although grazing permits are limited to ten years, they may be renewed. By custom, a fairly secure tenure of individual users has been established. This security, together with grazing fees which are low relative to private leases, has had the result that the number of grazing permits issued to an individual enterprise increases the value of the permittee's home ranch.

The second remedy is used in the United States with respect to

---

W. H. Osgood, *Increasing the Recovery of Petroleum* (New York: McGraw-Hill Book Co., 1930), 2 vols.

Northcutt Ely, *Oil Conservation through Interstate Agreement* (Washington: Govt. Print. Off., 1933), 398 pp. (U. S. Federal Oil Conservation Board).

M. W. Watkins, *Oil; Stabilization of Conservation* (New York: Harper and Brothers, Publishers, 1937), 269 pp.

U. S. National Resources Committee, Energy Resource Committee, *Energy Resources and National Policy*. (Washington: Govt. Print. Off., 1939), 435 pp.

Consult also articles and statistics in the *Minerals Yearbook*, U. S. Bureau of Mines and the *Oil and Gas Journal*.

[6] R. E. Hardwicke, "The Rule of Capture and Its Implications as Applied to Oil and Gas," *Texas Law Review*, Vol. 13, No. 4, June 1935, pp. 391–422.

[7] Ground water flowing in defined channels, in contrast to percolating ground water but similar to surface flow, is subject to riparian and appropriation rights as modified by the rule of "reasonable beneficial use." For details see Patricia McBride Bartz, with a foreword by S. V. Ciriacy-Wantrup, *Ground Water in California, The Present State of Our Knowledge* (Berkeley: University of California, College of Agriculture, Agricultural Experiment Station, September, 1949), 67 pp. (Giannini Foundation Ground Water Studies No. 1). Processed.

[8] Ten years later, 58 grazing districts contained 264.6 million acres of land, of which 133 million acres were vacant, unappropriated, unreserved public land and 8.59 million acres other public land. Twenty-two thousand operators grazed 10.77 million head of livestock.

wildlife. The Constitution has vested the people as a whole with ownership; both state and federal governments act as trustees for the people. Utilization of nonmigratory upland game and inland fisheries is controlled in great detail by state laws. Utilization of migratory waterfowl is controlled by federal law.

In contrast, ownership of wildlife outside territorial waters remains undefined. The resulting depletion by users from all nations is to some extent counteracted by existing international treaties which in certain areas (North Atlantic, Northwest Pacific, Baltic) regulate the utilization of whales, salmon, herring, and fur seal. The control of these remaining fugitive wildlife resources will be considered later, when the international tools of conservation policy are discussed.

Instability of Property Rights

Property rights may be well defined, but they may be unstable over time. Insecurity of tenure of this type is another important cause of economic uncertainty and resource depletion.

The following cases are examples of instability of property rights. A tenant who doubts whether his lease will be renewed will find it profitable to mine his soil, although soil-conservation practices may be quite economical under more stable tenure. Farmers or ranchers who own their land but who are obligated to pay high interest and amortization charges may fear they will be dispossessed by their creditors in times of drought or economic depression; they have no more interest in soil conservation than a tenant with a short lease. In periods of political change when stringent government regulations, confiscatory taxes, or outright socialization threaten a sudden and radical redistribution of property rights, all resource users will hesitate to make investments with deferred yield.

In all these cases, resource users will adopt a utilization plan which they themselves would regard as wasteful depletion if their property rights were more stable. From the standpoint of social economics, likewise, this type of depletion is wasteful. It is not a part of the concept, the social purpose, of property institutions.

Nor is the present structure of these institutions necessary for their concept. Rather, such depletion indicates that the structure has become inadequate to realize the concept.

Instability of property rights is such an important cause of resource depletion that remedial measures warrant a more detailed consideration. This will be undertaken when the derived property institutions (tenancy, credit, and taxation) are discussed. At this point in our broad survey of property rights it is more appropriate to mention a somewhat different and in a sense opposite aspect of the problem—rigidity of property rights.

Instability versus Rigidity of Property Rights

Emphasis on instability of property rights as a cause of wasteful resource depletion does not mean that absolute stability of property rights over time is desirable for conservation or, if it were, actually obtainable.

Historically speaking, the distribution of the individual strands of the bundle of property rights is different not merely for different societies—depending on their whole culture of which tenure institutions are only a part—but also for different periods in the life span of a given society. Such changes are necessary because the concept, the social purpose of property rights, changes over time. The structure, the legal body of property rights, must be adapted accordingly.

Discrepancies between concept and structure due to inflexibility of property rights create stresses and strains which are just as undesirable for conservation as those, mentioned above, created by instability. The results are sudden and radical redistributions of property rights. Expectation of such redistributions, as we saw earlier, leads to increases in uncertainty allowance and to resource depletion. Gradual, evolutionary changes in the distribution of property rights among the various participants in resource tenure are necessary from the standpoint of conservation, as from many other standpoints. This necessity is emphasized by the results of imbalanced property rights.

Imbalance of Property Rights

"Imbalance of property rights" will be understood in this study in a specific economic sense: property rights will be called "imbalanced" if they lead to such a distribution of revenues and costs from resource utilization among the members of a social group that the agent responsible for conservation decisions is not interested in taking *all* of these revenues and costs into account. Generally, such an agent will consider only those revenues which accrue to him (rather than to somebody else) and those costs which he (rather than somebody else) is required to pay. Although revenues and costs incident on others may be functionally related to a planning agent's decisions, such revenues and costs have no influence upon these decisions. The difference between the *incidence* of revenues and costs (on the basis of property rights) and their *allocation* (to functionally related use rates) and the consequences of this difference upon conservation decisions are of paramount significance in the social economics of conservation (chapter 17).

How property rights influence the incidence of revenues and costs in resource utilization and the importance of this influence for the conservation decisions of private planning agents are best shown by a few examples.

A farmer may find it profitable to deplete cheap foothill land and move somewhere else. The resulting runoff and soil erosion may do great damage through washing or deposition on valuable valley land and through siltation of water reservoirs. These costs may be far greater than the revenues obtained from the depleted foothill farm. Still, property rights usually are such that these costs have no influence on the utilization plan to which they are related. Public agencies may do similar damage through improper construction or drainage of highways.

A lumber company which finds a system of exploitative forestry economical is not legally responsible for the private and public costs of stranded workers, damaged watersheds, and loss of recreational opportunities.

In many areas private industry and urban communities are

still within their property rights when untreated wastes are "economically" disposed of in streams, lakes, or ground water, or near ocean beaches. The damage to others is not a cost which need be considered by those who cause pollution of water and air.

Depletion of scenic resources is commonly a cost borne by the general public and not by the many private resource users who cause the depletion.

On the other hand, laws regarding private property rarely grant support by the public or specific beneficiaries to those forest owners who realize the stake of the community in sustained-yield forestry, to those industries which take proper precautions against water and air pollution, and to those farmers who take pride in a well-kept landscape free of gullies, billboards, and rural slums.

Many more examples could be cited for cases in which property rights lead to such an incidence of revenues and costs from resource utilization that some revenues and costs (in extreme cases most of them) are not taken into account in the utilization plan which gives rise to them. In terms of social economics, distortions of the utilization plan result. The word "distortion" is justified because, as will be shown in chapter 17, one condition (among others) for the socially most desirable utilization plan is that *all* revenues and costs functionally related to given rates of use be considered.

Some might argue that such consideration is restricted by the effects of property rights; in other words, the distortions noted need not be corrected but should be regarded as a part of the concept, the social purpose of property rights. Admittedly, the concept of property rights varies greatly between different societies. In modern democracies, at least, such an argument would not be tenable. In these societies, property rights are regarded as instruments to make the self-interest of individuals serve the welfare of all. In terms of this ideal, imbalance in the above sense is contrary to the concept of property as a social institution. In these societies, therefore, such an imbalance may be interpreted also as an imbalance between the concept and the structure of property institutions.

Remedies for Imbalance of Property Rights

There are various ways for reducing this imbalance through changes in the structure of property rights. Only a short sketch of these various ways is attempted here because they are more properly discussed as tools of conservation policy in chapter 19.

One way to reduce the difference between incidence and (in terms of social accounting) proper allocation of revenues and costs from resource utilization is more effective application of the principle of compensation, both by perfecting existing legal instruments (civil law covering damage) and by devising new ones (government subsidies to conservation).

Another way is prohibition of those kinds of resource uses in which incidence and proper allocation of revenues and costs is particularly unbalanced. Examples for this approach are zoning restrictions and nuisance-abatement ordinances.

A third way is not the negative one of prohibiting certain uses but the positive one of requiring certain practices. Examples are state and local regulations requiring treatment of industrial and urban wastes in order to prevent water and air pollution, and some land-use ordinances by organized districts in the fields of flood control, irrigation, soil conservation, and range management.

CHAPTER 11

TENANCY

Meaning of Tenancy

Of the three derived property institutions, tenancy will be discussed first because its relations to conservation are fairly simple, as compared with credit and taxation.

The term "tenancy," as we know, refers to the relations between the owner and the user of resources—that is, to those strands of the bundle of property rights which the owner surrenders to the user.

Some students of law define "ownership" exclusively on the basis of rights to use.[1] Theoretically, a clear distinction between owner and user is thereby rendered impossible. Practically, at a particular time and place, property institutions set owner and user clearly apart. Roman as well as common law makes, between owner and user, important distinctions that have economic consequences. These consequences are significant enough to justify

[1] "The most important rights in land are those to which we apply the term ownership, involving, within limitations more or less wide, the idea of rights in some particular person or persons (the owner or owners) to use the land according to his or their pleasure, and to demand that others refrain from such use, the word ownership being used without reference to time, greater or less duration of the rights involved."—H. T. Tiffany, *The Law of Real Property* (3d ed.; Chicago: Callaghan and Co., 1939), Vol. 1, Sec. 2, p. 4.

For different views see W. W. Cook, "Ownership and Possession," *Encyclopedia of the Social Sciences*, Vol. 11, pp. 521–525.

a distinction between ownership and usership in the economics of conservation.

In the case of ownership, tenure includes rights of access to the resources and rights to a privileged position in the distribution of returns from resource utilization without time limit. It comprises also the rights to dispose of access rights and distribution privileges and to impose regulations upon those in whose favor such rights are disposed of.

In the case of usership alone (tenancy), on the other hand, tenure is confined to the bundle of rights that are surrendered by the owner for a limited period. Furthermore, the user is restricted in rights of disposal. He is greatly affected, not only by the owner's actual use of disposal rights, but by the mere possibility that the owner may use them.

Effects of Tenancy upon Conservation Decisions

In our problem, interest is centered on the conservation decisions of planning agents. The effects of tenancy need be considered, therefore, only from the standpoint of the parties who make conservation decisions. Division of the decision-making functions between owner and tenant is common. In such cases tenancy interests us in its effects upon the owner's *and* the tenant's conservation decisions. On the other hand, if the tenant has no functions with respect to conservation decisions—for example, some croppers on cotton plantations—tenancy needs no particular attention here.

Tenancy affects the state of conservation mainly through instability, through incidence of revenues and costs on owner and user, through fixed regressive rents influencing individual time preference, and through lessening the results of imperfections in the markets for physical, personal, and money assets. Each of these four effects of tenancy will be considered in turn. In order to avoid repetition later, we will also consider some aspects of conservation policy applicable to these effects of tenancy. Such consideration is confined here to the tools of conservation policy. The criteria and objectives of conservation policy will be discussed in chapters 17 and 18.

Instability of Tenancy

Uncertainty allowance on account of instability of tenancy is not necessarily determined by legal status but may be determined by the customs and traditions affecting subjective uncertainty of the planning agent about his tenure. In the United States, relations between owner and tenant of agricultural land are frequently based on a simple oral contract extending over one year. This highly insecure legal status is not necessarily reflected in a high allowance for uncertainty in the utilization plan. Local custom and tradition known and respected by both owner and tenant may make such a tenure quite definite and secure. There is scarcely any social institution in which local customs and traditions play a more important role than in tenancy. Statistics about past occupancy are significant only if past occupancy is reflected in expectations about future occupancy.

However, customs and traditions cannot quickly be established where they do not exist. Thus, in order to decrease uncertainty allowance on account of short, informal leases, emphasis must be placed on the general adoption of longer lease contracts. Lease contracts should contain clauses that require each party to notify the other, a sufficient time before expiration, whether and under what conditions renewal is desired. The latter may be partly accomplished through compensation for disturbance. A lease extending over three to five years with one-year notice of renewal would mean a considerable improvement of present leasing customs in many parts of the United States. In other countries, such leases would still be considered short, and the period of advance notice insufficient.[2]

On the other hand, it appears futile to strive for adoption of longer leases while both owners and tenants find the shorter ones to their advantage. Both parties may be reluctant to tie their hands for a long period because of general economic uncertainties and

[2] In central Europe, for instance, lease contracts for public and private holdings cover usually 12 to 24 years. Many contracts extend over longer periods. In the United States statutory limits are placed on the period of lease—in California, for instance, 15 years for farmland. Some large landowners in California—for example, the Southern Pacific Railroad Company—use a 5-year contract.

because of uncertainty about frictions between themselves arising from the execution of the lease. Reduction of these two causes of short leases and short occupancy appears as a prerequisite for increasing the use of longer lease contracts. A sliding-scale cash rent varying with prices—rather than the common fixed cash or share rent—may go far toward decreasing the reluctance of owners and tenants to enter into long leases because of general economic uncertainties. (Other advantages of sliding-scale rents will be discussed in following sections.)

An equitable determination of rent and a detailed, written lease setting forth clearly the rights and duties of both parties helps to avoid misunderstanding and frictions. One may determine an equitable rent by carefully computing the contributions made by both parties toward the net revenues of the leased property.[3]

Incidence on Owner and Tenant

The broad social significance of the problem of incidence versus proper allocation was mentioned earlier. At this point we are interested in how the incidence of revenues and costs on owner and tenant of resources affects conservation decisions.

If all expected revenues and costs functionally related to the actions of a tenant were incident on him, he would have no reason to alter the utilization plan because of the fact that he is not the owner.[4] In theory, the utilization plan would be the same even under short leases as under ownership. In reality, this result cannot be expected, but it may be approached through fuller compensation for deferred revenues and costs at the time of severance of tenancy.[5]

---

[3] Cf. "Empirical Method for Determining the Division of Farm Income," proposed by R. L. Adams and W. H. Smith, Jr. in *Farm Tenancy in California and Methods of Leasing* (Berkeley: 1941), pp. 40–43. (California Agr. Exp. Sta. Bull. 655.)

[4] Revenues and costs functionally related to actions of the planning agent but expected when his tenure is ended must, of course, be discounted to the date on which his tenancy ends; and proper allowance for uncertainty must be made.

[5] In England the principle of compensation was recognized in the Landlords and Tenant Act of 1851 and in rates established by the Agricultural Holdings Act of 1875 and made compulsory in 1883. Modifications were also made from time to time in later agricultural holding acts. In Germany, customary but well-defined rates of compensation for application of manure, fertilizer, plowing, seeding, etc., are in use. Customary rates of depreciation of these (and of permanent) improvements are employed.

Sometimes lease contracts, law, or local custom require the tenant to surrender a flow resource in the same condition in which he took over. If future flow is increased or decreased through his actions, he is compensated or must compensate the owner, at rates again fixed by contract, law, or local custom. Sometimes, the contracting parties agree that certain improvements will be made without special compensation. Compensation is then implicit in the terms of the lease and appears in a lower rent or in assistance by the owner through labor and materials. Local custom and the terms of the lease may also permit damage to the resource without special compensation to the owner. In an area where depletion is common, the tenant is rarely made responsible for the depleting effects of his actions upon use rates after his tenure has expired. Elsewhere, compensation to the owner for depletion may be implicit in the lease. In California, if land that was a number of years in alfalfa is rented for the use of truck crops (a rotation frequently involving change of tenant as well as crops) the owner expects a decrease in fertility and takes this into account in higher rents for the truck crops.[6] In renting stock resources, of course, decrease of the stock is agreed upon by both parties, and the owner is compensated by the payment of a depletion allowance which, strictly speaking, must be differentiated from other portions of the total rent payment.[7]

Clearly, the problem of compensation, like that of duration of leases, must be viewed in the light of local conditions and customs. In a country with such variable economic conditions as the United States, a rigid enforcement of detailed legal provisions relating to duration of lease and compensation may not be the most effective approach. Instead, formulation of general rules in state land tenancy acts may be suggested, with administration of these rules left to courts of tenancy arbitration, composed of owners and tenants with an experienced judge as chairman.[8]

[6] Such a decrease in fertility is not a *necessary* consequence—for example, if truck crops are fertilized or the soil otherwise brought into a better state of cultivation. (S. V. Ciriacy-Wantrup, "Economics of Joint Costs in Agriculture," *Journal of Farm Economics*, Vol. 23, No. 4, November 1941, pp. 771–818.)

[7] Cf. L. C. Gray, "Rent under the Assumption of Exhaustibility," *Quarterly Journal of Economics*, Vol. 28, May 1914, pp. 466–489.

[8] England has had a Land Tenure Act since 1851, Sweden since 1907, Germany

Owners and tenants should both be free to appeal to these courts in any question involving interpretation or execution of the lease. The mere existence of such a court would reduce the uncertainty caused by instability of tenancy. The courts could also draw up and recommend types of lease contracts suitable for their district and could educate owners and tenants in other ways. Many frictions between landlord and tenant and many expensive and slow suits before the regular courts could be avoided.

One condition for tenancy courts to act beneficially in the proper allocation of revenues and costs between owner and tenant is more knowledge about the deferred portions of revenues and costs. What are the deferred benefits of liming, terracing, fertilizing, and the like? What are the deferred costs of current "savings" in repair of fences, buildings, and so on? The agricultural experiment stations may well give more attention to the time distribution of the results from various practices.

Fixed Regressive Rents

Why fixed and regressive charges influence individual time preference and why and under what conditions they influence conservation decisions were explained in chapter 7. We concluded that, from the standpoint of conservation, fixity and regressiveness of rents (with respect to income) are undesirable. As noted, rents are often regressive because demand for small properties by prospective tenants is greater relative to potentialities of money income than the demand for large properties. The causes are imperfections in the markets for money loans, limitations in managerial capacity (partly caused by unequal educational opportunities), spatial and occupational immobility of operators, and extramarket revenues.

A well-developed system of tenancy along the lines just discussed is in itself a safeguard against the effects of imperfections in the loan market. How these imperfections themselves can be mitigated will be considered in the next chapter. Other policies

since 1920, the Netherlands since 1937. In all these countries, experience with special courts of arbitration has been excellent. For a detailed discussion of the provisions of the British act consult Aubrey J. Spencer, *The Agricultural Holdings Act, 1923* (8th ed.; London: Stevens and Sons, Ltd., 1931), 326 pp.

which are of interest in this connection relate to training and retraining facilities for planning agents, to subsidies for locational and occupational movements, and to assistance in a new location or occupation. Funds expended for this purpose may be made repayable after a certain period.

Fixity of rents can be avoided if cash rents are expressed as a percentage of net revenues.[9] However, such rents are rare because of difficulties in ascertaining all costs. An approximation is a rent variable with gross revenues. The various sliding-scale cash and share rents now used in California agriculture are of this type. For example, cash rents for dairy farms vary with monthly butter production and prices; pasture rents vary with beef or lamb quotations; share rents for cotton, raisins, prunes, and other crops vary with prices. The U.S. Forest Service charges grazing fees to its tenants which vary with livestock prices as a percentage of a base rate per cow-month or sheep-month.

The worst effects of fixity of rents can be avoided by variable-payment plans. Under such plans a surplus over the fixed normal rent is accumulated in periods of high net incomes, and payments are reduced below the normal rent if subnormal production (for example, because of drought) or subnormal prices (as during a depression) lead to a dangerous fall of net incomes. Variable-payment plans have actually been employed in cases of fixed interest and amortization payments on farm loans (chapter 12). There appears no reason why the same principle could not be used for fixed rents.

From the standpoint of variability of rents, the common fixed share rent is superior to the common fixed cash rent. The common share rent, however, may affect the state of conservation in a way that is of interest at this point.

In agriculture, the landlord may belong to a low-income group. Under this condition a decrease in the value of the landlord's share, because of a fall in product prices, increases his time preference. Many share leases allow the landlord considerable influence upon the utilization plan because he participates in the

[9] In California costs of harvesting vegetables are frequently deducted from gross, and rent is figured as a percentage (for example 15 per cent) of this "net."

risk to a much greater degree than in fixed cash leases. This influence will result in depletion under the assumed conditions. Evidently, therefore, a fixed cash rent rather than a share rent is better suited to mitigate effects of time preference upon the state of conservation if the landlord belongs to a low-income group and if his influence upon the production plan of the tenant is great. In all other cases a share rent or, still better, a cash rent variable with income is preferable from the standpoint of effects upon conservation decisions.

According to some students of tenancy, the crop share lease in contrast to the stock share lease prevents tenants from adopting the optimum state of conservation. Cause and effect may be confused if the crop share lease itself is made responsible for land use, which sometimes is of a lower state of conservation than land use under a stock share lease. These two lease forms adapt themselves to land-use systems based on crops and livestock, respectively. This type of adaptation is more common and more logical than an adaptation of land-use systems to lease forms. In California, for instance, where many land-use systems coexist in the same general region, lease forms vary greatly from one farm to the next and change with changes in land use over time. From the standpoint of conservation, the crop share lease as such is not particularly objectionable.

Imperfections in the Markets for Assets

It was suggested in the preceding pages that a system of tenancy can be devised in which the time distribution of resource use would not be greatly different for the tenant than for the owner. Under such a system, conservation policy may even look with favor upon tenancy because it tends to counteract imperfections of markets for natural resources, managerial ability, labor, loans, and equipment.[10] The reasons for these imperfections were

[10] The importance of tenancy in relation to imperfections in the markets for natural resources, managerial ability, labor, and loans is well known. The importance of tenure arrangements for the equipment market is illustrated by coöperative machinery associations, which own heavy machinery needed in the building of terraces and other soil-conservation practices and rent this machinery to individual members.

explained in chapter 7. They interfere with realizing the optimum state of conservation, although in a particular situation they may tend toward either depletion or conservation (relative to the same situation with perfect markets).

It would not be realistic, however, to expect that improvements in the system of tenancy prevailing in the United States will be made to an extent sufficient for preferring tenancy to ownership from the standpoint of conservation. Thus, attempts [11] to transform tenants into owners, although often costly, are in line with the objectives of conservation policy provided that new owners are set up in economical units and without too high and fixed charges for interest and amortization (chapter 12). However, the process of eliminating tenancy is slow at best. Conservation policy cannot be content with it. As long as tenancy exists, improvements in the direction suggested here are of equal if not greater importance. Improvement of the tenancy system does not conflict with attempts to increase the proportion of owners. Both policies may be pursued at the same time.

The Government as Landlord

In the United States, the federal government is by far the most important landlord in terms of land area owned and numbers of tenants. In the states west of the 100th meridian, the policies of the federal government on its grazing lands are an important factor in the local economy.

At first glance it does not seem logical to give special consideration to tenancy of public land in the private economics of conservation: could one not expect that a public landlord has the best possible system of tenancy with respect to the conservation of resources?

It is true that public landlords generally try to take resource conservation into account. They foster stability of tenancy through assigning priorities for grazing permits, provide for compensation in cases of inexhausted improvements at the termination of

---

[11] The Bankhead-Jones Tenant Farms Purchase Act of 1937 [50 *U. S. Stat. at L.* (1937), p. 522] is an example. This act was superseded in 1946 by the Farmers Home Administration Act [60 *U. S. Stat. at L.* (1946), p. 1062].

the lease, coöperate with tenants in such improvements, employ rents (grazing fees) variable with livestock prices, and insist on lease clauses and regulations designed to avoid overgrazing.

Public landlords, however, are in a peculiar position relative to their tenants: The public servants who represent the landlord (officials of the U.S. Forest Service, Grazing Service, and so on) are subject to a considerable amount of political pressure by the tenants through both local organizations and representatives in Congress. Although these public servants of the landlord are not directly employed by the tenants, the latter have frequently much more influence upon them than have the great majority of taxpayers who are no less employers but whose individual interest is smaller and who are frequently less informed.

For the sake of argument, we will make the assumption that the tenants of public lands are as much interested in resource conservation as the landlord. Still, the peculiar relation between public landlord and his tenants becomes relevant for the social optimum in the state of conservation if competitive uses exist, for example, if use for livestock competes with the use for wildlife and recreation. At a certain degree of use, depletion of the resource can be avoided only by limiting some of the competing uses. It is only natural that the political influence of the tenants will be exercised strongly in favor of curtailing other uses—for example, in favor of reducing deer herds and of more stringent trespass regulations. In order to avoid an imbalance of tenure rights—in this case to the disadvantage of many users and of the great majority of owners (the taxpayers)—a greater influence of other interests than that of tenants in the leasing of public lands would seem desirable. This influence may be provided through local, state, and national advisory boards.

Such boards exist for Taylor grazing districts and for national forests. A widening of the administrative functions of these boards—besides their advisory function—is advocated by livestock interests. Such a change may not be objectionable provided *all* users of public land, including wildlife and recreation interests, received proper representation. Public servants would be strengthened in their attempts to avoid an imbalance of tenure rights if conflicts of interest are frankly discussed and as far as possible settled within advisory boards.

At present, the composition of Taylor grazing district advisory boards is as follows: Ten elected cattlemen and sheepmen in number roughly proportional to the number of cattle and sheep permittees in the district, one elected "free use" representative (permittees with a limited number of livestock kept for domestic purposes), and one appointed wildlife representative. The state advisory boards are composed in most cases of a cattleman and a sheepman elected from each district board. Similarly, the national advisory board consists of a cattleman and a sheepman elected by each state advisory board. Members of advisory boards in the national forests are all elected by livestock interests. At present, therefore, representation of various interests on the advisory boards is strongly one-sided.

Another device to smooth conflicts between landlords and tenants of public lands in matters concerning resource conservation is an appeal board. Such a board has been established on the national level for the U.S. Forest Service as an independent fact-finding agency making recommendations to the secretary of agriculture. If an appellant asks in writing for a hearing, that hearing must be granted. Experience with the board is not yet sufficient to say what influence it has actually had. Discussion of a conflict in resource conservation before an agency outside of the Forest Service may help to decrease the danger of political pressure.

CHAPTER 12

CREDIT

Credit in Relation to Tenure

The credit system is the second important economic institution which is derived from that of property. True, the creditor has no tenure rights in resources by the mere existence of debt. However, in most modern societies such rights are voluntarily surrendered by debtors in the loan contract, and many statutes relative to foreclosure and moratoria contain restrictions on use rights in order to protect creditors.[1]

The credit system, similar to tenancy, is related to the state of conservation mainly through uncertainty allowance, through fixed and regressive charges influencing individual time preference, through lessening imperfections in the markets for physical and personal assets, and through conservation clauses in the loan contract.

[1] For example, in California it is unlawful for any person in possession of real property to do any act of injury to real property during foreclosure. (*California Code of Civil Procedure* [1941], chap. 3, sec. 745, p. 384.) The trustor or mortgagor maintains and repairs the property and does such other things as the court deems just and equitable for the protection of the security. No sale of personal property (with certain exclusions) used in connection with operation is allowed until after sale of real property. (*Cal. Stats.* [1941], chap. 204, p. 1263; especially sec. 5, p. 1265; sec. 6, p. 1265; sec. 17, p. 1269; and sec. 19, pp. 1269–70.)

Under United States statute, creditors may supervise the farming operations of a farmer. The cost of such supervision may partly be borne by the farmer. This supervision is in addition to that of the conciliation commission. (52 *U. S. Stat. at L.* [1938], p. 84.)

Credit and Uncertainty of Tenure

The credit system may create uncertainties of tenure that resemble the uncertainties discussed in the preceding chapter. If the owner's equity is small in relation to his creditors', then ownership does not guarantee a secure tenure. Even if the owner succeeds in fulfilling the loan contract, the creditors may recall the loan before it actually becomes delinquent, in order to safeguard their equity. This, at least, has occurred with private loans given on a strictly business basis—that is, excluding loans given by family members and by public institutions. Inability to fulfill the contract and recall of the loan usually happen at a time when refinancing or liquidation of assets through sale is difficult or impossible. The result is bankruptcy and foreclosure. This threat cannot fail to influence the utilization plan. Allowance for these uncertainties must result in depletion. In other words, the owner may find it advisable to liquidate through depletion as much as possible of the present value of his resources before foreclosure prevents him from saving his equity. This possibility emphasizes the conclusions we have already reached with respect to transforming tenants into owners as a measure of conservation policy: if such a change of tenure status is connected with high indebtedness, the owner's allowance for uncertainty of tenure may be no less than the tenant's.

These undesirable effects of the credit system upon the state of conservation may be reduced by setting up legal machinery for debt adjustments without foreclosure. Such machinery may be set up like the special tenancy courts suggested in the preceding chapter or may be combined with them. In the latter case they could properly be called tenure courts.

Credit and Fixed Regressive Charges

Generally, payments connected with credit (interest, amortization) are fixed charges. Fixed charges, as we saw in chapter 7, tend to discourage conservation under imperfect markets for assets. Furthermore, fixed charges connected with the credit system are usually regressive with income. Debtors in low-income groups pay higher interest rates than debtors with larger incomes because such risks are considered greater by lenders [2] and because the cost of administering smaller loans is higher per unit. The relation between regressiveness of fixed charges and conservation was likewise discussed in chapter 7.

The most radical measure to prevent the influence of fixed and regressive interest and amortization payments upon the state of conservation is to limit, or to prohibit by law, indebtedness in low-income groups of resource users—for instance, on all farms of family size.[3] This would exclude one important possibility of lessening the influence of imperfections in the markets for land, labor, and management. Such lessening generally tends toward conservation, as will be illustrated in more detail below. From the standpoint of conservation, therefore, prohibition or curtailment of credit may defeat its own purpose. Furthermore, it would exclude one economic group from the dangers and opportunities affecting all others. Such a protected group may then not be able to function as a healthy part of a society in constant change.

A less radical measure is to make interest and amortization payments variable with the debtor's income. This may be done in two ways: by variable-payment plans or by relating interest and amortization payments more directly to the prices of products. Variable-payment plans were previously suggested in connection with rent payments. With respect to interest and amortization for loans, such plans are authorized under Title I of the Bankhead-Jones Act, mentioned earlier. The secretary of agriculture might offer purchasers of farms assisted by the act a payment plan continuing over forty years "under which a surplus above the required payment will be collected in periods of above

[2] This may not be justified. If debtors with small incomes have family enterprises, risks may be in reality smaller. Through accepting smaller labor incomes and working longer and harder, the family may be able to earn enough for interest payments during a depression. This is not possible for debtors with larger enterprises and correspondingly larger incomes (during periods of prosperity) because their incomes are derived largely from profits and interest from invested capital, not from the labor performed by the debtor's family.

[3] This was a feature of the Peasant Estate Inheritance Act (during the Hitler regime) in Germany, through which all farmers owning an acreage up to 310 acres were prevented from contracting any long-term debts. Short-term production credits at regulated rates were permissible.

normal production or prices and employed to reduce payments below the required payments in periods of subnormal production and prices." [4] Contracts between the Bureau of Reclamation and irrigation districts frequently contain some provision for adjusting the annual repayment of construction costs in accordance with changes of farm income. In some high-risk farming areas, variable-payment plans are becoming popular in private debtor-creditor relations as well. In the Great Plains area, for example, about 5,000 farms have changed hands under variable-payment plans; also many loan companies in this area have introduced the principle of forbearance.[5]

Similar results may be obtained by expressing the loan and, therefore, amortization and interest in terms of the main product or products. In agriculture, for instance, loans may be expressed in terms of oranges, corn, wheat, cotton, or in units of a composite of grains and livestock or fruits and vegetables; in industry, loans may be expressed in terms of coal, pig iron, or any composite of manufactured products.[6] This, of course, decreases fluctuations of income only if they are caused by price fluctuations of products not connected with the individual debtor's rate of production. Fluctuations of income in periods of upswing and downswing in resource industries operating under price competition are of this type, in contrast to income fluctuations caused by weather.[7] A decrease in the amplitude of income fluctuations, brought about by making interest and amortization payments flexible, is not

[4] In the Farmers Home Administration Act of 1946, the successor of the Bankhead-Jones Act, payments can be reduced only to the extent that surpluses above the required payments have been accumulated in previous years. Flexibility was thus reduced somewhat.

[5] Elmer Starch, "The Future of the Great Plains Reappraised," *Journal of Farm Economics*, Vol. 31, No. 4, Pt. 2, November 1949, p. 924.

[6] In the period of violent inflationary and deflationary price movement in central Europe following the First World War, lenders as well as borrowers favored loans to agriculture on the basis of rye, the most important grain crop. Loans to heavy industries were based on coal. In these cases, of course, effects upon time preference were not the motive.

[7] S. V. Ciriacy-Wantrup, "Major Economic Forces Affecting Agriculture, with Particular Reference to California," *Hilgardia*, Vol. 18, No. 1, December 1947, pp. 1–76. (University of California, College of Agriculture, Giannini Foundation of Agricultural Economics, Paper 121).

S. V. Ciriacy-Wantrup, *Booms, Depressions, and the Farmer* (Berkeley: 1948), 24 pp. (Calif. Agr. Exp. Sta. Circ. 376).

only desirable because of its effect upon the borrower's utilization plan, but is also favorable for the lender: the need for moratoria, refinancing, and foreclosures in periods of depression is reduced; and the purchasing power of interest income is protected in periods of prosperity.

A third measure is special public credit aid to low-income groups of resource users—for example, through the loan department of a Farm Security Administration. This possibility leads us to a more general point—namely, that the credit system can become an important aid to conservation.

Credit as an Important Aid to Conservation

The two foregoing relations between credit and conservation—through uncertainty of tenure and through fixed and regressive charges—are undesirable both in the private and in the social economics of conservation. However, as has been indicated, these relations are, to some extent, avoidable. Further, they are more than balanced by the relations between credit and conservation which are economically and socially desirable. These relations were previously implied in chapter 7: A well-functioning credit system reduces the effects (upon the state of conservation) of imperfections in the markets for physical and personal assets and of changes in time preference. Such a credit system is an effective aid in realizing the optimum state of conservation.

Theoretically, such aid may facilitate either conservation or depletion; this would depend on the direction in which the above imperfections tend to deflect the state of conservation from the optimum in the absence of such aid. Practically, however, such aid usually tends in the direction of conservation. In other words, a well-functioning credit system is generally more important to those resource users who want to borrow funds for conservation than to those who want to borrow funds for depletion or lend funds obtained through depletion.

As we have seen in chapter 7, the credit system usually functions with considerable imperfections. A reduction of these imperfections, therefore, warrants special attention from the standpoint of conservation.

Imperfections in the Credit System

The credit system may be imperfect because of monopolistic conditions, because of greater risks on small and scattered farms, or because of institutional lags.

If markets for loans are imperfect because of monopolistic conditions, it may be desirable to increase competition through public banking institutions and through assistance to coöperative efforts of borrowers, for example, coöperatives of the Raiffeisen and Schulze-Delitzsch type.

Public credit, preferably in the form of credit insurance, is also desirable if credits to many small and scattered enterprises are involved. The justification of public assistance in order to facilitate the pooling and spreading of credit risks was discussed earlier. Such pooling and spreading, plus better access to credit funds, might enable a public forest credit system to offer long-term credit to small and medium forest enterprises at lower rates than private lenders could afford without assistance through public credit insurance.

A distinction should be made between public lending institutions that compete with and act as a yardstick for private lending, and ones that serve borrowers regarded as submarginal by the private credit system. In the United States, for instance, the Farm Credit Administration falls into the former and the Farm Security Administration and other emergency agencies into the latter group.

Many imperfections in the credit field are caused by institutional lags. In a new country, the allowance for uncertainty taken into account in interest rates may not be reduced by lenders as much as is justified by the increase in economic stability brought about by the maturing process. Uncertainties of new industries and new enterprises, likewise, may not be allowed for in accordance with their development. Methods of appraising assets may have become inadequate and obsolete. Estimates of costs for administering loan contracts may need revision—for instance, because of better transportation facilities in rural areas. Insistence on certain types of collateral may defeat its own purpose in view of changes in physical and economic conditions. On Western ranges, for instance, the number of livestock grazed is often the collateral for loans and, therefore, the basis for loan rationing. Lenders object to having the number reduced even if such a reduction would prevent a decrease in the value of the collateral. In the past, when overgrazing was less of a problem, such a lending practice may have been justified by the instability of the grazing industry. But it is obsolete now; and a change would eliminate a serious cause of privately and socially uneconomic depletion.

Lack of private credit is sometimes caused by this type of inertia. For example, it took some time and the successful activities of public land banks before insurance companies regarded agriculture as a safe and profitable investment. As yet, little lending is done by banks and insurance companies in forestry. In this field the definition of "improved property," which restricts lending in the field of forestry, needs adaptation.

Conservation Clauses in Credit Agreements

Finally, one aspect of the relations between credit and conservation may be mentioned which could become of considerable practical importance. As pointed out at the beginning of this chapter, the creditor is frequently granted influence upon the debtor's utilization plan through clauses in the loan contract or even by statute. If the government is the creditor, the loan contract may become the instrument of regulations which for political and constitutional reasons cannot be applied by themselves, at least for the time being. In forestry, for example, socially desirable silvicultural practices may be introduced as a part of a public forest credit system. In agriculture, private and public banks may make proper soil conservation a condition for lending. The Federal Land Bank of St. Louis and some private banks in Oklahoma [8] and in the Middle West are doing some valuable pioneering in this field. The same opportunities exist in international lending through the International Bank created by the Bretton Woods agreement.

---

[8] The Oklahoma Bankers Association has been active in calling attention of its members, of farmers, and of the public to the importance of soil-conservation practices.

CHAPTER 13

TAXATION

Significance of the Tax System in Resource Conservation

In resource conservation the practical significance of taxation is probably even greater than that of the two economic institutions just discussed. Like the other derived property institutions, the tax system has highly significant but unintended, unrecognized, and socially undesirable effects upon conservation decisions of private planning agents. In this sense taxation is frequently an important obstacle for conservation policy. On the other hand, the tax system can be employed more easily and effectively as a tool of conservation policy than can the tenancy or credit system. As before, in order to avoid repetition, we will discuss the possibilities of using taxation as a tool of conservation policy together with the effects of taxation in private economics. The objectives and criteria of conservation policy will be discussed later.

The effects of taxation upon conservation decisions are far more complex than those of tenancy and credit. These complexities have some similarity to problems already encountered in connection with prices. Our approach, therefore, will follow a similar line of reasoning.

The Necessity of Simplifying Assumptions

Any attempt to make general statements about the effects of taxation upon conservation decisions faces the same difficulties we encountered with prices: when new taxes are levied or existing ones changed, we need to know how these changes are distributed over time, and how interrelations between use rates in different intervals are affected as compared with the pretax [1] situation.[2] Changes of taxes will be interpreted broadly—that is, will include changes of tax rates, of methods of assessment, of modes of payment, and similar changes.

As with prices, we will make simplifying assumptions with respect to the expected time distribution of tax changes and with respect to the effects of these changes upon complementarity, competitiveness, and independence of differently dated use rates: we will assume that a given tax change is expected to last indefinitely or, at least, over the whole planning period; we will also assume that interrelations between use rates are not affected. Before proceeding on the basis of these two assumptions it is well to consider problems that arise if they are *not* fulfilled.

Little needs to be said about cases in which the *second* assumption is not fulfilled. Taxes are generally not imposed specifically on "conserving" or "depleting" services. If they are, interrelations between use rates will be affected in much the same way as by increases in the prices of such services. The effects were discussed earlier. Taxes may also be imposed on "conserving" or "depleting" products. The effects of this situation in multiple-product enterprises like agriculture are similar to those caused by price decreases. This problem, likewise, has been discussed in the chapter on prices.

More important at this point are cases in which the *first* assumption is not fulfilled—that is, cases in which variations of taxes during the planning period need consideration.

---

[1] The term "pretax" will henceforth refer to the situation which existed before the *change* in taxes that is considered took place.

[2] The use that is made of the proceeds of the tax by the tax authority is also important—for instance, if the proceeds are used to subsidize conservation. The effects of such a subsidy should be regarded, however, as a separate influence to be analyzed as such and not confused with the direct effects of taxation.

Variations of Taxes over Time

Variations of taxes during the planning period may be expected by private planning agents as an intended part of governmental tax policy; for example, the government may imply or even publicly announce that a given change of taxes will be limited in duration and that the pretax situation will be restored at a certain date. Sometimes, planning agents may expect variations of taxes on the basis of extrapolating their past experience, or on the basis of important present events—for example, a war or a depression—which are known to lead to great changes of fiscal needs.

If a given change in taxes is not expected to last indefinitely (or over the whole planning period), rates of use tend to be redistributed over time in such a way that the tax base is increased in those planning intervals in which taxes are more favorable.[3] Redistribution goes on as long as the discounted savings in taxes are greater than the decrease in present net revenues that would have been caused by such redistribution under pretax conditions. Conservation or depletion of any degree may result according to the time distribution of the tax changes.

Such redistribution is of practical importance in conservation policy: Through appropriate variations of taxes over time, with or without public announcement of intended later changes, the government can encourage depletion (for example, through tax reduction on stock resources during a national emergency) or conservation (for example, through increasing fees for hunting and fishing licenses).

Taxation may even go so far as to force productive services (management, labor, equipment) entirely out of certain fields of resource use for some time. Productive services will shift into less heavily taxed sectors of the economy or into leisure. The result is conservation in the more heavily taxed sectors provided special

[3] If successive changes of taxes are *not* expected by private resource users (on the basis of government announcement, extrapolation of past experience, or otherwise), influences in the same direction as those considered in the text may still be observed or even intended by the government. In such cases we would have conservation and depletion *ex post*—brought about through the processes of selection (chap. 6).

cases [4] of complementarity of use rates and of shifts from conserving to depleting services and products are excluded. To some extent, such effects of taxation are limited by constitutional provisions.

Progression and Regression in Taxation

Some relations between conservation and progression or regression were considered in connection with the analysis of time preference; these relations were not confined to taxation. We are now interested in another relation between conservation and progression [5] (or regression). This relation is mentioned at this point because of its similarity to the effects of variations of taxes over time.

Most private utilization plans are characterized by variations of the tax *base* over time. Under these conditions, if a progressive tax is imposed, or the progression of existing taxes is increased, use rates will be redistributed in such a way that the tax base in different planning intervals becomes more equal. Again, redistribution proceeds as long as the discounted savings in taxes are greater than the decrease in present net revenues that would have been caused by such a redistribution under pretax conditions. Conservation or depletion of any degree may result, depending on the pretax variations of the tax base over time.

Regressive taxes have a tendency to accentuate, not to level, differentials (between intervals) in the tax base. Accordingly, their effect upon the state of conservation tends in the opposite direction from that of progressive taxes.

Here, also, the redistribution is of practical importance for conservation policy: Frequently, particularly in the utilization of *stock* resources, the tax base for individual firms is much smaller in the more remote than in the immediate future; this situation results from the effects of cumulative use upon costs and

[4] Such special cases may occur, for example, in mining when temporary interruption of operations, especially for repair and maintenance, makes later resumption of operations uneconomical. Similar special cases occur in forestry and will be considered in connection with yield taxes.

[5] Tax exemptions in the lower brackets may be regarded as a special case of progression.

revenues. Under these conditions progressiveness in taxation will cause a redistribution of use rates toward the future, and regressiveness a redistribution toward the present. This result strengthens the conclusion reached in chapter 7: progression in taxation tends toward conservation and regression toward depletion as compared with a situation in which the same aggregate amount of taxes is collected without progression or regression.

On the other hand, in the utilization of *flow* resources the tax base may be higher in the more remote than in the immediate future. For example, up to a certain limit, cutting of trees becomes more profitable the further it is postponed in an immature forest. In such cases progression of income taxes may influence operations to start sooner than would be desirable for conservation. If mature stands are being cut, progression of income taxes may delay liquidation. In this case, earlier liquidation would result in conservation (chapter 4, footnote 3). However, in forestry, substitution of the capital-gains tax (which is not progressive in the United States) for the income tax (which is highly progressive) works against such depleting influences (see p. 175).

Tax-saving Certificates

The relation between progression of taxes and conservation should be considered in connection with a recent proposal to use taxation for inducing farmers to accumulate cash reserves in regions with great variability of yield—for example, in the Great Plains wheat belt.[6] Under this proposal a farmer could save up to one-half of taxable income in high-income years tax-free by purchasing "tax saving certificates" that would be issued for this purpose by the United States Treasury. Redemption of the certificates would be at the option of the purchaser—presumably in low-income years—and the proceeds would be taxed as income in the year of redemption.

In view of the existing progression of United States income taxes, the proposed scheme would tend to make variations of

[6] E. Lloyd Barber and Philip J. Thair, "Institutional Methods of Meeting Weather Uncertainty in the Great Plains," *Journal of Farm Economics*, Vol. 32, No. 3, August 1950, pp. 391–410.

the tax base—that is, year-to-year variation of income caused by variations of yield *or* prices *or* costs—more profitable than it is now. It would become more profitable to "mine" the soil in years of good moisture or prices by decreasing diversification in crop rotation in favor of wheat, by plowing up grasslands, and by extending wheat cultivation into areas climatically or topographically not suitable for permanent wheat cultivation. Exploitative, temporary farming by speculators interested in avoiding the progression of income taxes would receive a premium. Tax-saving certificates would be a good investment for anyone in the higher-income brackets confronted with income variations over time.

True, cash reserves would be increased by the proposed scheme; but a portion of these reserves would represent merely a monetization of fertility reserves.[7] In other words, the decrease in progression of income taxes would result in soil depletion. This danger by itself may not make the proposal socially undesirable. However, it should be carefully considered. An accentuation of existing tendencies toward soil depletion and toward speculative wheat farming may more than offset—in its effects upon the long-run stability of the area—an increase of cash reserves.

Preceding Effects Exist Regardless of Types of Taxes

The effects of factors just discussed (shifts between practices and products, variations of taxes over time, progression and regression of taxes) hold for all types of taxes (income taxes, property taxes, death taxes, yield taxes, and so on). For example, income taxes, which are frequently regarded as neutral with respect to the time distribution of use rates, may have decisive effects upon the state of conservation through progression or regression and through shifts of productive services (management, labor, equipment) into or out of certain fields of resource utilization.

[7] Wheat stored on the farm in good years would result in similar savings of income taxes. But this form of reserves is more costly because of storage costs and losses, less attractive in years of high prices, and generally less suitable for a "cut out—get out" operation. The proposal cautioned against here would make wheat storage less attractive as an economic alternative.

Before we turn to the analysis of various types of taxes, a short remark on the problem of capitalization of taxes should be added. Capitalization of taxes is significant for analysis from the standpoint of the distribution of the tax burden among taxpayers—that is, from the standpoint of income allocation. However, capitalization does not affect allocation of productive services over time. In any event, its possible influence on conservation decisions is small and indirect. From our standpoint, therefore, capitalization of taxes need be considered only in passing.

Income and Profit Taxes

Taxes on income and profit are taxes on current net revenues, in contrast to recurrent (annual) taxes on the present value of the whole stream of expected future net revenues—as in property taxes.

In studies of taxation the view is expressed again and again that income and profit taxes will not be shifted because the planning agent has no incentive to change the pretax system of utilization. Aside from problems of definition (see the following paragraph), the common view would be correct in time economics only if income and profit taxes are proportional (neither progressive nor regressive)—and under the assumptions discussed above that (1) they are also expected to be constant over time, (2) that no shifts of productive services into leisure or into less heavily taxed employments take place, and (3) that the familiar income effects upon individual time preference under imperfect markets for assets can be disregarded. True, under these assumptions, *proportional* taxes on current net revenues do not offer any incentive to redistribute current net revenues and, as a necessary prerequisite, rates of use over time; in other words, they are neutral with respect to the state of conservation. However, great care must be exercised in probing where these assumptions are fulfilled in actuality.

Somewhat different effects of income and profit taxes upon resource conservation are connected with the legal definition of "income" and "profits" for tax purposes. This definition frequently does not coincide with the meaning of net revenues in economic theory. From our standpoint, two discrepancies are of special interest. They refer to the segregation of current costs from investment and to the employment of depletion allowances or depletion accounts. These two problems are highly important for conservation of both flow and stock resources in the United States.

Segregation of Current Costs from Investment

In agriculture and forestry, many expenses for permanent improvements can be charged, for income-tax purposes, to current costs of production. Partly this can be done in conformity with income-tax laws; partly it is done because segregation of investment from current costs, as required by law, is difficult. Expenses of this kind are, for example, land leveling and terracing, reseeding of permanent pastures, enlargement of drainage and irrigation systems, planting of trees, fencing, and gradual improvement of roads, buildings, and equipment. Such expenses are investments resulting in conservation. At a certain level of income-tax rates it is economical to avoid or to evade income taxes by making such investments. This may be the case even if—in accordance with our previous assumption—income-tax rates are expected to remain the same in the future, and if the effects of progression and regression are disregarded. The explanation is the following: Taxpayers can sell their improved properties at a profit and are taxed on the latter only on the basis of their capital gains. The capital-gains tax is considerably lower than the income tax in the higher income brackets.[8] Thus, a high income tax in conjunction with a lower capital-gains tax may result in conservation.

A somewhat different "substitution" of the capital-gains for the income tax exists in forestry. In the United States, timber owners who cut their own timber can treat profits from timber operations as a capital gain which can be offset—with some exceptions—by capital losses in the same or in any other operation of the taxpayer.[9] This provision puts owner-operators on the same

[8] For individuals, the capital-gains tax is 50 per cent of reportable net gain; reportable net gain is 50 per cent of actual net gain if capital assets have been held over 6 months. A taxpayer may pay income tax on reportable net gains if it is less.

[9] 58 *U. S. Stat. at L.* (1944), p. 21.

basis as timber owners who sell stumpage outright. Generally this provision will result in conservation: depleting effects of progression are prevented (p. 172); the tax load is reduced and, therefore, effects upon individual time preference are less.

Depletion Allowances

There are other types of special income-tax considerations for some resource industries which result in depletion. An important example may illustrate this.

In the United States, producers of stock resources (metal ores, nonmetallic minerals, oil, gas, coal, and the like) receive important special consideration—that is, as compared with such industries as manufacturing, agriculture, transportation, the service industries—in income and profit taxation. First, during the "development stage" only expenditures in excess of current receipts need to be charged to capital account. In oil and gas production this provision is especially liberal: it allows offsetting development costs not only by receipts from the enterprise being developed but also by receipts of the taxpayer from other sources. Second, depletion allowance is permitted which is not based on development costs and is not influenced by the fact that development costs may be fully recovered. Such depletion allowance is either a 50 per cent reduction of the net operating income or a statutory percentage of gross income—whichever is the lesser—throughout the life of the enterprise.[10]

What are the results of these provisions in the income-tax law? Exploration for *and* production become more profitable relative to other activities than prices of products and productive services would bring about under a more equalitarian income tax. More capital will be attracted into the field, prices will be lower, and consumption higher. This in itself need not necessarily mean depletion according to our definition in chapter 4. However, lower

[10] The statutory percentage of gross income is considerable. At present it is 27½ per cent for oil and gas; 23 per cent for sulfur; 15 per cent for metal ores and nonmetallic minerals; 5 per cent for coal.—Title 26, *U. S. Code* (Internal Revenue Code), Sec. 114 (b), (3) and (4), p. 2579.

Some changes in this legislation are under discussion. However, on previous occasions, the Congress has been opposed to any reduction of depletion allowances.

prices of the primary product will discourage the utilization of scrap, reduce the incentive to develop and employ conserving processes in production, refining, and consumption, decrease imports, and delay the development of substitutes. For a given country, the available stock will be depleted faster, and ultimate recovery will be lower.

At present, as during the Second World War, these depleting influences in the regular income tax are accentuated by certain provisions in the Excess Profits Tax Act of 1950 which exempts many stock resources altogether, and reduces progression greatly for others. As we know, progression in income taxes on stock resources generally tends toward conservation. By the same act, direct government subsidies to depletion are largely exempted from income taxation; this situation also prevailed during the Second World War.

Accelerated depletion of domestic stock resources may be desirable *during* a war. As a condition existing for an extended period of time *prior* to a war, such depletion may become dangerous: domestic resources may be depleted to such an extent that requirements cannot be met in a later actual war emergency even if depletion is accelerated to the utmost.

In times of peace or semipeace, subsidies to the depletion of domestic stock resources—directly and in the form of special income-tax provisions—might better be used to stimulate exploration without depletion, to encourage research (in substitutes, in utilizing scrap, in the economies of consumption), to finance stand-by capacity, and to facilitate imports of foreign resources without injury to the capacity of domestic producers. More will be said about this when the national-security aspects of conserving domestic stock resources are discussed in relation to imports of foreign resources.

Taxes on the Value of Property

Property taxes are among the most important taxes in resource utilization. Besides the general-property tax, more special taxes on the value of land, mines, buildings, and equipment are of this type. In the United States the early adoption of the general-property tax left little room for these special taxes, which have always been the backbone of resource taxes in European countries; but with the differentiation of property and the growing evasion of taxes on personal property, the general-property tax has become largely a special tax on physical assets (natural resources, improvements, and equipment).

For the purpose of analysis, the value of physical assets is assumed to be identical with the sum of discounted future net revenues which these assets are expected to yield. This assumption is in conformity with the goal of most tax authorities although, in actuality, it is only approximately correct. The effects of deviation from this goal because of inadequate methods of assessment will be indicated later. Between concepts as used by tax authorities and concepts as used in economic theory, there are also the same sort of discrepancies as there are with taxes on current net revenues. However, the following analysis of the effects of property taxes holds regardless of the particular type of computation—for example, by the famous Hoskold formula [11]—that is employed by tax authorities.

Recurrent (annual) taxes on the present value of resources may be regarded as a special type of taxes on net revenues. If present value is the sum of discounted future net revenues, then in each interval (year) in which the tax is paid, net revenues of all future intervals are taxed. The further, therefore, net revenues are distant from the present, the more often they are subject to the tax. This provides an incentive to redistribute net revenues in the direction of the present, in order to reduce the number of times they are taxed. This process continues as long as discounted savings in tax payments are larger than the decrease in present net revenues that would have occurred with such redistribution under pretax conditions. Redistribution of net revenues in the direction of the present can be accomplished only through redistribution of use rates in the same direction. This means depletion. Storage may be regarded as part of the productive process;

---

[11] Charles H. Baxter and Roland D. Parks, *Mine Examination and Valuation* (2d ed.; Houghton, Michigan: Michigan College of Mining and Technology, 1939), p. 126.

anticipation and postponement of revenues and costs through credit operations may bring about obvious modifications; the extent of these modifications, however, is limited. Property taxes, therefore, affect the utilization plan in much the same way as the interest rate.

A simple numerical example will illustrate these effects so important in practice. Let us assume two alternative time distributions of net revenues (A and B) for a utilization plan extending over four years as follows:

Series	t_1	t_2	t_3	t_4
A	$1,000	$1,000	$1,000	$1,000
B	$1,500	$1,500	$ 500	$ 400

At an interest rate of 4 per cent, the present value of series A is $3,775.09; that of series B, $3,760.18.[12] The planning agent will, therefore, choose series A which, under identical technology and economic conditions, results from a more conservative time distribution of use rates. If the two underlying time distributions of use rates are the only alternatives, that distribution on which series A is based may be called the optimum state of conservation.

Now let us assume that a property tax of 3 per cent is imposed. To obtain the expected value of the property tax in each year during the whole period of utilization, present values as of the first day of each year are computed:

Series	t_1	t_2	t_3	t_4
A	$3,775.09	$2,886.09	$1,961.54	$1,000.00
B	$3,760.18	$2,350.59	$ 884.62	$ 400.00

The expected property tax for each year is then:

Series	t_1	t_2	t_3	t_4
A	$113.25	$86.58	$58.85	$30.00
B	$112.81	$70.52	$26.54	$12.00

12 For simplicity all current net revenues are discounted as of the first day of each year.

The sum of the expected property taxes discounted at 4 per cent to the present is $277.58 for series A and $215.82 for series B. Deducting these values from the pretax present value of the two net revenue series, we obtain $3,497.51 for series A and $3,544.36 for series B. Through imposition of the property tax, therefore, the lower state of conservation underlying series B becomes the optimum state of conservation. The planning agent will change his utilization plan accordingly. This change means depletion.

The illustration is so chosen that the lower state of conservation becomes the optimum state through imposition of a property tax. Other examples which could be used would show merely that a property tax decreases the present value of a conservative utilization plan relatively more than that of a less conservative plan.[13] Such a decrease, however, is sufficient to bring about a change in the optimum state of conservation if the alternatives in the time distribution of use rates and the resulting present values are numerous, as can be assumed in economic theory. In theory, a certain state of conservation is always marginal—that is, alternatives with a higher or lower degree of conservation are close at hand.[14] In economic reality the alternatives confronting the planning agent are discontinuous, as in our example.

[13] If, for instance, series B is chosen so as to give ultimate cumulative net revenues equal to series A (t_1, $1,500.00; t_2, $1,500.00; t_3, $500.00; t_4, $500.00), the less conservative distribution of use rates is the optimum one before imposition of the tax, as long as a positive interest rate is used. Through imposition of the tax its superiority becomes greater. The relative present values of series B (series A = 100) are 101.96 before the tax is imposed, 103.58 afterward.

[14] Fairchild's statement on this point is confined to a special case: "From its very nature, the property tax favors a use which yields an early income. Of course, this effect is controlling only in the case of those properties which are on the margin between use for a deferred yield and use for annual return." Fairchild compares two time patterns of use, one characterized by deferred yield (forestry) and the other by annual yield (grazing).—Fred Rogers Fairchild and associates, *Forest Taxation in the United States* (Washington: Govt. Print. Off., 1935), p. 46. (U. S. Department of Agriculture. Miscellaneous Publication 218.)

The effect of the property tax "is controlling," not only if deferred yield and annual yield are close alternatives, but also if different time patterns of annual yield are alternatives. In forestry a choice between numerous different deferred-yield cycles must be made, and the cycles planned need not be constant over time. In grazing, many time patterns of use between reckless overgrazing and extreme restriction are of economic importance. Confinement to constant use rates or to strictly periodic variations of use rates over time simplifies too greatly the economic problems involved.

The depleting effects of property taxes are sometimes denied on the ground that a property tax will be capitalized and will merely induce the sale of the taxed property to financially stronger speculators, without affecting the time pattern of use.[15] Property taxes are, without doubt, easily capitalized when the taxed property is sold; such a sale may or may not be forced by the tax. But the new owner feels the same economic inducement under the tax to change the pretax time pattern of use as the old owner. As mentioned before, we must distinguish sharply between the effects of the tax with respect to the tax burden and with respect to the distribution of use rates over time. In this study, interest is focused on the latter effects, which are not altered by the capitalization of taxes.

Deliberate Use of Depleting Effects of Property Taxes

Sometimes, the depleting effects of property taxes are desired by tax authorities. The two most important examples are attempts to check speculation in resources withheld from use and attempts to change methods and practices of use. Both cases need some clarification in the light of the previous analysis.

Sometimes the private optimum in the state of conservation is characterized by extreme curtailment of immediate use in the expectation of high capital gains later. Vacant city lots, large land holdings in young, rapidly growing countries, and deposits of minerals under monopolistic conditions (chapter 14) are examples.

In such cases, taxes based on current utilization—net revenue taxes or yield taxes—would produce no tax receipts in the beginning or only low ones and would not interfere with a time distribution of use rates that may be regarded by tax authorities as socially undesirable. Property taxes, on the other hand, produce tax receipts immediately and change the time distribution of use rates in the direction desired.

These desirable results of the property tax depend on an im-

---

[15] C. H. Sargant, *Urban Rating* (London and New York: Longmans, Green, and Co., 1890), pp. 145 ff.

William Smart, *Taxation of Land Values and the Single Tax* (Glasgow: J. MacLehose and Sons, 1900), pp. 95 ff.

portant assumption: it must be economically possible to develop the resource after imposition of the tax. A too heavy property tax may actually prevent resource development and merely lead to tax delinquency.[16] In this connection it may be noted again that property taxes—like all other taxes—can be used (within constitutional limits) to force productive services into less heavily taxed fields or into leisure. This means conservation in the more heavily taxed fields—if special cases of complementarity of use rates are excluded (footnote 4).

Property taxes on resources already used, but not used in conformity with existing economic potentialities, are sometimes recommended because of their alleged educational value in forcing planning agents to adopt better methods and practices or to sell their property to others who will.[17]

Taxation does not change imperfections of knowledge, or individual nonmonetary goals of utilization such as prestige, class distinction, seclusion, and simplicity of management. Property taxes change the utilization plans of *all* planning agents in the direction of depletion, but leave unaltered imperfections in knowledge and differences in goals. The burden of property taxes is, of course, greater (in proportion to actual current net revenues) for properties the utilization of which falls short of realizing the highest potential in current net revenues. Whether this leads eventually to a sale of these properties depends on the financial strength of their owners. Other owners, who use resources in conformity with economic potentialities, but who are financially weaker, may be affected first.

At best, the property tax is a slow and uncertain instrument for bringing about the adoption of better methods and practices in resource utilization. If taxation is to be used for this purpose, a tax directly on *potential* current net revenues would be preferable

[16] Fear of this effect led the Minnesota legislature in 1941 to change the property tax on taconite to a yield tax. Taconite cannot compete with the high-grade ores of the Mesabi Range as long as they last. The intention of the tax change was to encourage experimentation with taconite. Depletion of taconite would help in the conservation of the Mesabi Range ores.

[17] This is the basis for Aereboe's often repeated doctrine of the social benefits from land taxes.—Friedrich Aereboe, *Agrarpolitik* (Berlin: P. Parey, 1928), 619 pp., particularly pp. 315–327.

to a property tax: it would not cause a general tendency towards depletion; and it would still impose an especially heavy burden on the utilization plans that are based on imperfect knowledge or are not oriented towards maximization of present monetary net revenues. More direct public action through education or regulation may produce results more quickly, surely, and economically than taxation (chapter 19).

Depleting Effects of Inequalities in Assessment

In the United States the depleting effects of property taxes are accentuated by certain aspects of assessment. According to numerous studies of the general property tax, the ratio of assessed value to present value tends to decrease as present value increases.[18] The tendency of regression in taxation—in this case through assessment—to favor depletion has been explained earlier.

Another inequality in assessment, also of great importance in the economics of conservation, is the overassessing of properties in a low state of conservation relative to properties in a high state. Such overassessment has been studied particularly with respect to cutover forest land with poor natural regeneration,[19] but it applies also to eroded farmland and overgrazed ranges. A vicious circle is set in motion: inequalities of assessment encourage depletion most on those properties which are already relatively more depleted. This effect is often increased by tax regression.

---

18 Eric Englund, *Assessment and Equalization of Farm and City Real Estate in Kansas* (Manhattan, 1924), 70 pp. (Kansas Agr. Exp. Sta. Bull. 232.)

H. W. Yount, *Farm Taxes and Assessments in Massachusetts* (Amherst, 1927), pp. 88–120. (Massachusetts Agr. Exp. Sta. Bull. 235.)

W. H. Dreesen, *A Study in the Ratios of Assessed Values to Sales Values of Real Property in Oregon.* (Corvallis, 1928), 45 p. (Oregon Agr. Exp. Sta. Bull. 233.)

H. D. Simpson, *The Tax Situation in Illinois* (Chicago, 1929), 104 pp. (Northwestern University, Institute for Research in Land Economics and Public Utilities, Studies in Public Finance. Research Monograph No. 1).

Whitney Coombs, *Taxation of Farm Property* (Washington, 1930), 75 pp. (U. S. Department of Agriculture Tech. Bull. 172).

L. P. Gabbard, *Inequalities in Taxation of Farm Lands and City Property Due to Scope and Method of Assessment* (College Station, 1932), 28 pp. (Texas Agr. Exp. Sta. Bull. 458).

19 F. R. Fairchild, *op. cit.*, Part 4, pp. 77–150.

Deferred Reassessment of Property Taxes

In some cases tax authorities attempt to make property taxes less discouraging for conservation. This can be done by leaving property taxes unchanged, at least for some time, if the value of the property is increased through the efforts of the owner.

One such attempt was recently reported as part of the program of the present regime in North China: [20] land having an increased normal yield because of conservation practices or other improvements is not to be reassessed (in terms of standard *mou*) until the fourth year after the improvement.

Such a deferment in reassessment should be differentiated from the deferment of tax *payments*, which is a general feature of yield taxes in cases of deferred yield. This latter deferment will be discussed presently under yield taxes. Deferment of a change of property taxes mitigates but does not change fundamentally the characteristic tendency of these taxes to favor depletion.

Death Taxes

In the modern Western world death taxes are based on the present value of the estate.[21] They are called "estate taxes" if the tax is imposed regardless of the number of beneficiaries or their relationship to the testator. They are called "inheritance taxes" if taxes are imposed on the shares of the beneficiaries; in this case exemptions and tax rates usually differ with the relationship between each beneficiary and the testator.

If the utilization plan extended over several generations, death taxes would have the same inherent tendency to cause depletion as recurrently payable property taxes. It may be assumed, however, that only one payment of death taxes is usually taken into account. If this assumption is granted, death taxes affect the optimum state of conservation in the same way as a proportional tax on future net revenues.[22] According to the previous analysis of

---

[20] *Foreign Agriculture*, Vol. 14, No. 2, February 1950, p. 36.

[21] In some parts of the world death taxes are lump-sum taxes, or they are based on some physical expression of present wealth, such as the number of livestock owned.

[22] Proportional with respect to future net revenues in different intervals for the same estate. With respect to present value of different estates, death taxes are usually strongly progressive.

net-revenue taxes, therefore, death taxes are neutral with respect to the optimum state of conservation under perfect markets for services. However, under imperfect markets death taxes may cause depletion because of their influence upon the time preference of testator and beneficiaries: if the testator expects that a portion of his estate will be turned over to the government and will not directly benefit his heirs, his motives for investment may be weakened, or he may be induced to disinvest.[23] Under imperfect markets for loans and for other productive services, disinvestment may take place in the form of depletion. The beneficiaries, in turn, are faced with payment of a tax that usually far exceeds current net revenues from the inherited assets. If they have no liquid reserves, and if markets for loans and other important productive services are imperfect, they have to liquidate a portion of the inherited assets through depletion in order to pay the tax. This type of depletion is common in forestry and agriculture.

To avoid these effects of death taxes, payments should be extended in installments over such a period that current net revenues can meet current payments.[24] In Great Britain, for instance, death taxes on forest holdings need be paid only at the time the timber is actually cut. For the sake of formal completeness, lowering of death taxes and avoidance of progression may be mentioned,[25] but these remedies appear socially questionable on grounds not connected with the economics of conservation.

---

[23] The opposite effect is theoretically possible, but it can be assumed so rare, especially under progressive tax rates, that it may be neglected.

[24] The extreme in this respect would be a transformation of death taxes into income taxes. On this point, see:

Tibor Barna, "The Burden of Death Duties in Terms of an Annual Tax," *Review of Economic Studies*, Vol. 9, No. 1, November 1941, pp. 28–39.

N. Kaldor, "The Income Burden of Capital Taxes," *Review of Economic Studies*, Vol. 9, No. 2, Summer, 1942, pp. 138–151.

[25] The National Committee on Inheritance Taxation recommends a maximum of 15 per cent of present value. See National Committee on Inheritance Taxation, *Report to the National Conference on Estate and Inheritance Taxation Held at New Orleans, Louisiana, November 10, 1925* (Washington: 1925), 118 pp. (Reprint containing model succession and estate tax laws.)

Yield Taxes

Under yield taxes are grouped a great variety of resource taxes. Some of them are imposed on units of physical production, regardless of sales; others are imposed on units of the physical product sold, regardless of home consumption. Severance taxes are frequently of these types. Still other yield taxes are imposed on the value of product sold (ad-valorem taxes, gross-revenue taxes).

Yield taxes may be regarded as deductions from current net revenues in relation to use rates (provided net revenues are positive and home consumption is disregarded in those cases in which yield taxes are imposed only on the portion of production that is sold). Thus, if yield taxes are proportional (that is, neither progressive nor regressive) they offer no incentive to redistribute use rates over time. They are neutral with respect to the state of conservation. In this, they are superior to property taxes. However, for this result the same assumptions need to be made which were discussed earlier for taxes on current net revenues.

Like all other taxes, yield taxes may be used to force productive services temporarily into less heavily taxed fields or into leisure. Generally, as we know, such shifts mean conservation for the more heavily taxed resource—at least from the standpoint of the tax authority. However, some exceptions to this result were previously noted for mining.

Similar exceptions may now be noted for forestry: In forestry a yield tax may prevent important conservation practices. It may be important for conservation, for example, to thin a young stand in order to increase future growth of the remaining stand, or to cut inferior species which would reproduce an inferior stand for future forests, if left when the superior species are harvested. A yield tax on thinnings and on inferior species may make such practices uneconomical. Net-revenue taxes and property taxes would not have these specific effects. However, a yield-tax law may be drawn up so as to avoid such effects. In Oregon the first 25,000 board feet cut are exempt from taxation.

Yield taxes are superior to most other taxes with respect to economy and accuracy of assessment and administration. They are inferior to net-revenue taxes from the standpoint of ability to pay and social justice, because costs are not taken into account. They are less flexible in periods of income changes caused by factors other than yield.

Lump-Sum Taxes

Lump-sum taxes are levies of fixed amount on individual enterprises, payable currently without regard to use rates, or payable only if use rates are greater than zero. The pretax optimum state of conservation is, therefore, altered if it is economical to rearrange the utilization plan in such a way as to avoid lump-sum taxes in certain intervals altogether. Such rearrangement will occur as long as discounted savings in tax payments are greater than the decrease of present net revenues that would have been caused by such rearrangement under pretax conditions.

If lump-sum taxes are imposed without regard to use rates, the only possible method of avoiding them is to shorten the period of utilization. On the other hand, if lump-sum taxes are payable only if rates of use are greater than zero, one may find it economical to let rates drop to zero in certain intervals but to stay in business. The first type of lump-sum taxes causes depletion. Theoretically, the second type of lump-sum taxes may cause conservation or depletion depending on *when* rates of use drop to zero. In actuality, the effects will usually be depletion because it is generally more economical to let rates drop to zero in the more distant intervals of the utilization plan. Exceptions may occur, however; for example, if a cutover area is taken over with a view to growing timber. Even in this case, a yield tax or a net-revenue tax would be more effective in aiding conservation than a lump-sum tax.

Lump-sum taxes are inflexible and generally regressive. They may, therefore, lead to depletion for reasons which are already familiar. Moreover, this inflexibility and regression of lump-sum taxes may make them socially undesirable for reasons outside the field of conservation economics.

CHAPTER

14

MARKET FORM

Market Form and Conservation

Although property institutions have important relations to market form—they influence the number of market participants through patent laws, ownership of scarce resources, the systems of credit and taxation—it is not advisable to regard market form as a derived property institution. Many laws (Sherman Act, Agricultural Marketing acts, Fair Trade Practices acts), administrative agencies (Federal Trade Commission, Interstate Commerce Commission), customs (behavior of consumers in retail trade, leader-follower relations in pricing), mores (attitudes of a social group towards competition, regard by oligopolists for informal agreements), and special marketing arrangements (exchanges, auctions) have little to do with property institutions and decisively influence market form through effects upon the number of market participants and upon their market behavior.[1]

The form of the market in which products of resource users are sold and productive services bought is an important institutional force in resource utilization. Some relations between market form and conservation have already been encountered in this

[1] Emphasis in modern theory of market form on numbers (many or few) of market participants should not lead to the idea that there is a simple, direct relation between market form and numbers. The influence of numbers cannot be separated from that of economic institutions.

study. For example, the significance of imperfections in the markets for assets was emphasized in relation to time preference; these imperfections are not always connected with market form, but the latter play an important part.[2] When market form first entered our discussion, another relation to conservation was noted. This relation will now be explored in greater detail.

We noted that, under the market form defined as pure competition, product prices expected in different planning intervals are identical with marginal revenues expected for the same intervals. Under pure competition, prices can be regarded as independently given to individual resource users. We concluded that under pure competition, rates of use in different intervals were independent in revenues; thus interrelations of differently dated use rates through revenues could be neglected from the standpoint of their influence upon conservation decisions. For this reason we focused on the interrelation of differently dated use rates through costs.

On the other hand, we noted that, under market forms grouped together as "monopolistic conditions," interrelations of differently dated rates of use through revenues require an integrated production plan, influence conservation decisions, and therefore demand attention in the economic analysis of conservation. Such an analysis is the objective of the present chapter.

As we saw earlier, the terms "conservation" and "depletion" always imply a comparison between two different states of conservation. For the present purpose, the comparison will be between states of conservation under different market forms, assuming otherwise similar technological and economic conditions. Market forms and technological conditions are interrelated in many ways. Still, for economic analysis such a separation must be made.

In the last two decades the economic literature on market form and especially on monopolistic conditions has increased tremendously. There is no need to go into the many intriguing and controversial details of this literature. Our aims are rather

[2] It may be well to repeat that, in our terminology, an imperfect market includes, but is not confined to, monopolistic conditions. For example, a purely competitive market may be imperfect because of lack of information of buyers and sellers (chap. 7).

limited, namely, to understand, first, whether and how monopolistic conditions in resource use influence the time distribution of use rates in the direction of conservation or depletion and, second, to explore whether and how such influences are of interest from the standpoint of conservation policy.

The term "monopolistic conditions" is intended here to cover a number of market forms with greatly different economic characteristics (chapter 5). For example, in some respects pure monopoly differs more from Chamberlin's monopolistic competition or from oligopoly than from pure competition. A special volume would be needed to deal with all these different characteristics. However, the above and other monopolistic conditions (dyopoly, bilateral monopoly) have one characteristic in common: for individual planning agents, rates of use in different intervals are interrelated not only through costs but also through revenues. In this respect monopolistic conditions differ from pure competition. Thus, the effect of various market forms upon the time distribution of use rates may usefully be compared with that distribution which would prevail under pure competition.

Nearly all of the available literature on monopolistic conditions relates to simultaneous (instantaneous) economics. The results of these studies may be briefly summarized before we take up the influence of monopolistic conditions upon output in time economics.

Monopolistic Conditions and Rates of Use in Simultaneous Economics

Under a number of highly restrictive assumptions,[3] it is easy to show that the use rate under a pure monopoly is generally less than that of an "industry" [4] under pure competition; the

[3] These assumptions are discussed by Joan Robinson, *The Economics of Imperfect Competition* (London: Macmillan and Co., Ltd., 1933), 352 pp. Cf. particularly Book IV, chap. 11 (Comparisons of Monopoly and Competitive Output), pp. 143–154, and chap. 14 (Objections to the Comparisons), pp. 166–176.

[4] The difficulties of defining an "industry" cannot be discussed here in detail. For the purpose of our comparison, we mean by an "industry" all firms (one, several, or many) producing products which although they may be "differentiated" are substitutes for each other. The quantitative definition of substitutability depends on the purposes of the analysis. For our purpose a definition of "competitiveness in marginal revenues" similar to that employed in connection with resources and use rates may

same holds for pure monopsony and bilateral monopoly.[5] How much less is produced depends on the shape, whether concave or convex, of the demand and supply curves of the industry in question. If both curves are straight lines, the monopolist's use rate will be exactly half the rate of the industry under pure competition. The more convex the demand curve and the more concave the supply curve viewed from above, the closer monopoly output will be to competitive output.

Not only the shape of the demand and supply curves, but the possibilities of price discrimination, must be considered in tracing the effects of monopoly on output.[6] If the monopolist can discriminate perfectly in his purchases of productive services as well as in his sales of products, his marginal-revenue function becomes identical with the demand function; and his marginal-cost function becomes identical with the supply function of the industry under pure competition. Monopoly output will then be the same as the output of the competitive industry. The income distribution, of course, will be different in the absence of compensating fiscal measures. Perfect discrimination, however, is merely a theoretical possibility; in actual practice, monopoly output will be less than the output of the industry under pure competition.

However, there is a potentially more important influence which tends to make the traditional comparison between competitive and monopoly output unrealistic. This influence is sales promotion in the wider sense, including not only advertising but all other efforts to change the habits and tastes of consumers. It is frequently contended that sales promotion under pure monopoly is not profitable because the monopolist does not face competition. However, as previously mentioned (footnote 4), products of all

serve (chaps. 3 and 5). With respect to natural resources, the difficulties inherent in such a definition are generally less than for manufacturing industries. In a broader sense, all products can be regarded as competitive because they compete for the consumer's dollar and for available productive services (pp. 191–192).

[5] The assumptions relevant for bilateral monopoly are discussed by James N. Morgan, "Bilateral Monopoly and the Competitive Output," *Quarterly Journal of Economics*, Vol. 63, No. 3, August 1949.

[6] For a discussion of this aspect see A. C. Pigou, *The Economics of Welfare* (4th ed.; London: Macmillan and Co., Ltd., 1938), especially chaps. 16 and 17.

industries are substitutes for each other in the sense that they compete for the consumer's dollar. A monopolist, therefore, may find it profitable to promote his sales in order to attract customers from other industries. Such a shift in the demand function due to sales promotion may make the monopolist's output larger than that which would prevail under pure competition where sales promotion is not profitable.

Next, let us consider the output of an industry operating under monopolistic competition in Chamberlin's sense.[7] Under such competition individual firms are not of optimum size and have not reached the optimum degree of horizontal and vertical integration; the industry is usually burdened with large expenditures for sales promotion. If such an industry were transferred to monopoly control, it might be possible to reduce the marginal cost function of the monopoly below the supply function of the industry by reorganizing the industry with respect to size of firms, integration, and sales promotion.[8] Conceivably, therefore, rates of use may increase if an industry operating under monopolistic competition is transferred to the control of a pure monopoly.

On the other hand, as just indicated, the aggregate amount sold by an industry depends, in part, on the aggregate advertising outlays of that industry. To the extent, therefore, that advertising leads consumers to buy the (differentiated) product of an industry operating under monopolistic competition, rather than some other products, aggregate output may equal or exceed the output which would prevail under pure monopoly or even under pure competition.

In actual economic life, an industry operating under monopolistic competition is more likely to be transferred to monopoly control than an industry operating under pure competition. The same conditions that tend to uphold pure competition make it difficult for a monopolist to gain control. The government and

---

[7] Edward Chamberlin, *The Theory of Monopolistic Competition* (Cambridge: Harvard University Press, 1936), 225 pp.

[8] Generally, aggregate expenditures for sales promotion by an industry operating under monopolistic competition will probably be larger than such expenditures would be if the industry were transferred to the control of a pure monopoly, although such expenditures may not cease, as pointed out previously.

the public may be more tolerant of monopoly control, or may even encourage it (for example, in the form of a private cartel under government supervision), if the alternative is monopolistic rather than pure competition.

Our problem becomes even more complex when we attempt to compare aggregate output under oligopoly [9] with that under pure competition, monopolistic competition, and monopoly. Under oligopoly, advertising tends to make for larger aggregate output as compared with pure competition and, usually, with monopoly. Larger size of firms may permit more economical operations and a larger aggregate output as compared with monopolistic competition. However, these general tendencies are overshadowed by the necessity for all firms to take each other's actions and reactions into account in planning. The possible games of strategy which can be played between the oligopolists (or their coalitions) are so numerous that no summary of the outcome in terms of aggregate output relative to that under other market forms can be attempted here.

In conclusion, the often repeated statement that monopolistic conditions tend toward curtailment of output must be applied with caution even in simultaneous economics. This statement is true if monopoly output is compared with aggregate output of an industry under pure competition under highly restrictive assumptions. It is frequently not true if monopoly output is compared with aggregate output of an industry under monopolistic competition or under oligopoly. Furthermore, it may not be true if aggregate output under the latter two market forms is compared with that under pure competition. Let us see whether similar conclusions are reached if the tools of time economics are used.

---

[9] Dyopoly is not mentioned separately because it is rare in resource industries. With the increase in the number of firms, oligopoly shades into monopolistic competition and, further, pure competition. One may differentiate oligopoly from monopolistic competition on the basis of (1) tendency toward coalitions, (2) tendency toward equilibrium, (3) size of firms.

Monopolistic Conditions and Rates of Use in Time Economics

In order to make statements about the influence of monopolistic conditions on the time distribution of use rates, we have to know whether present use rates are complementary, competitive, or neutral with respect to later use rates through revenues. If these relations are competitive, we would expect that monopolistic conditions would tend toward conservation as compared with pure competition (where these relations are neutral from the standpoint of individual planning agents). If these relations are complementary, we would expect depletion.

As we know, differently dated use rates may be competitive through revenues if the resources themselves or their main products are durable (diamonds, other precious and common stones, metals, buildings, machinery, equipment, fertilizer over the period of gestation, grain over the period of storage), and if scrap from worn-out and obsolete products competes with the primary resources (secondhand bricks and lumber, and especially metal scrap). These conditions are fairly common. They are especially common in those industries in which monopolistic conditions prevail (nonferrous metals, diamonds, precious stones, fertilizers). In these industries, therefore, one would expect a more conservative use of resources than would prevail under pure competition. Indeed, the conservation argument is used in defense of monopolistic conditions sometimes by economists and quite commonly by interested parties, when the "wastes of competition" in resource utilization are decried.

What is the validity of this argument? According to our definitions of conservation and the above reasoning, this argument would be valid if the effects upon the aggregate output (weighted by time) of the industry by competitiveness of differently dated use rates through revenues are not offset or more than offset by the effects of aggregate expenditures for sales promotion. Under pure monopoly such an outcome is possible but not probable; [10]

[10] See footnote 8.

under monopolistic competition and under oligopoly it may not be unusual.

Sales promotion in itself has some bearing on the interrelations of differently dated use rates: present sales may be regarded as necessary to promote future sales—that is, in order to "hold the market" against the intrusion of others. Maintenance of sales even at a loss—for example, during depressions—may be practiced in order to facilitate future sales.[11] Whether such beliefs—that differently dated use rates are complementary rather than competitive through revenues—are justified or not does not alter their effectiveness in changing the time distribution of use rates in the direction of depletion, as compared with the state of conservation that would prevail under pure competition.

In conservation the effects of two economic forces—namely, interest and uncertainty—need special consideration. We must inquire, therefore, what influence monopolistic conditions have upon allowances for interest and uncertainty in production planning.

Interest rates, as we know, are generally regressive with size of the loan; the size of the loan varies usually with the size of the borrowing firm; likewise, imperfections in the credit market are usually less for large than for small borrowers. However, the relations between size of firms and market form are highly complex with respect to cause and effect. Size and monopolistic conditions are not always positively correlated. Under monopolistic competition, for example, size of firms may be small; under monopoly and oligopoly, firms are usually large. The last two market forms may, therefore, encourage conservation through lower interest rates and less imperfect loan markets.

It is sometimes contended that monopolistic conditions reduce the uncertainty of long-range investing. So far as this is true,

[11] Promotion of future sales is, of course, not confined to maintaining present sales at a loss. Under some conditions, large expenditures may be made to promote future sales without any present sales whatsoever. During the last war, the automobile industry and other producers of durable consumer goods made large expenditures for sales promotion—"There is a Ford in your future"—although no civilian cars were being produced.

monopolistic conditions would encourage conservation. Here again, differentiation between pure monopoly on the one hand and oligopoly and monopolistic competition on the other would seem relevant. Under the two latter market forms individual firms face important uncertainties—concerning the actions of their competitors—which are additional to the uncertainties faced by the monopolist and by the planning agent under pure competition. On the other hand, under oligopoly, agreements and quasi-agreements to reduce these uncertainties are common. Under such agreements—as under pure monopoly—it is likely that uncertainty allowance is less than under other market forms. This would encourage conservation.

We may conclude that, in time economics, retardation of output—like restriction of output in simultaneous economics—is not a *necessary* result of monopolistic conditions. This is particularly true if the time distribution of output under monopolistic competition and oligopoly—rather than pure monopoly [12]—is compared with that under pure competition. In real life, monopolistic conditions in resource industries are better described by monopolistic competition and oligopoly than by pure monopoly.

Monopolistic Conditions and Fugitive Resources

The familiar problems connected with indefiniteness of tenure and encountered in the utilization of fugitive resources are often not differentiated clearly enough from the relations between market forms and the state of conservation. This confusion exists because difficulties of defining tenure disappear if the resource as a whole or a definable portion of it is under unified control. However, unified control—for example, of an oil pool—

[12] For pure monopoly, without large expenditures for sales promotion and with competitiveness of differently dated use rates through revenues, we may agree with Hotelling's conclusion that ". . . monopolistic exploitation of an exhaustible asset is likely to be protracted immensely longer than competition would bring about or a maximizing of social value would require. This is simply a part of the general tendency for production to be retarded under monopoly."—Harold Hotelling, "The Economics of Exhaustible Resources," *Journal of Political Economy*, Vol. 39, No. 2, April 1931, p. 152. These conclusions are often repeated by writers who fail to take into account the necessary explicit and implicit assumptions.

leads to conservation not because the agent in control has any monopolistic powers—there may be many other pools—but because absence of the necessity of capturing the resource, before others capture it, influences the utilization plan. With fugitive resources, pure monopoly would be a *sufficient* but not a *necessary* condition for avoiding the effects of insecure tenure upon the state of conservation.

Monopolistic Conditions and Conservation Policy

The preceding pages tried to show that the relations between monopolistic conditions and conservation are more complex than is generally admitted. Public action aimed at changing market forms is never, therefore, a reliable tool of conservation policy—whatever advantages or disadvantages such action may have from other viewpoints. Let us assume, nevertheless, that such action is attempted.

In such a case, substitution of pure monopoly for oligopoly or for monopolistic competition is likely to encourage conservation. This kind of substitution, furthermore, may have advantages from other standpoints: it may be possible to rationalize an industry with respect to size of firms and sales promotion; it may be easier for the government to control output, prices, and profits of a monopoly than those of several or numerous firms under oligopoly or under monopolistic competition. On the other hand, there is the danger that the monopoly may become so strong economically and politically that its control will become difficult. Under oligopoly and monopolistic competition, economic and political power is divided. Usually, this division can be used by the government, without resort to totalitarian methods, to prevent concentration of economic and political power before it becomes a threat.

Under certain assumptions discussed later, it is possible to show that pure competition is one condition (among others) which must be fulfilled if private conservation decisions are to establish a state of conservation which may also be regarded as desirable from the standpoint of social economics (chapter 17). The traditional policy of the United States government is to safeguard or to reëstablish pure competition. This policy is backed by public opinion and embodied in the laws of the land.

Historically speaking, the result of this policy has been to prevent pure monopolies, or at least to hamper them considerably. In other countries such monopolies have come into existence in the form of integrated trusts, cartels, and corporative entities comprising whole industries—frequently sponsored and regulated by governments. However, it is equally true that the United States policy has not succeeded in substituting pure competition for such monopolies. As a rule, monopolistic competition and (bilateral) oligopoly has been the result. From the conservation standpoint, as shown above, this result may not be desirable.

The Sherman Act, the Federal Trade Commission, and other antitrust measures in the United States were not designed to aid in conservation. It is not advocated here that these measures be changed because of their possible negative effects in this special field. However, when the relevant choice is not between pure competition and monopoly but between a regulated monopoly on one side and oligopoly or monopolistic competition on the other, a more analytical, less emotional, attitude toward cartels and commodity agreements in the field of natural resources may sometimes be advisable. More will be said about this in the discussion of the effects of economic instability and the international tools of conservation policy.

CHAPTER

15

ECONOMIC INSTABILITY

Two Aspects of the Relations between Economic Instability and Resource Use

Commonly, the aspect of the relations between economic instability and resource use that is emphasized is the opposite of the one in which we are interested here: commonly, resource depletion is looked upon as the cause of economic and social instability.

In the long perspective, depletion of resources has been regarded a cause of the disintegration of societies and whole civilizations. In the New World this theory has been advanced to explain the fate of the Mayan civilization in ancient Yucatán. In the Old World similar hypotheses have been applied to the course of the Syriac civilization in what is now Iraq and of the branch of the Indic civilization in Ceylon. Historical data to test such hypotheses are rather meager; and the social sciences have not explored the available material sufficiently. Until this is done, definite conclusions are better avoided. It is not necessarily less plausible, in these and similar cases, that resource depletion has been the symptom rather than the cause of economic and social instability. In any event, the intricate connections between resource use and social relations must be more fully considered.

In the more recent past, depletion of individual resources has been responsible for economic and social instability in countries and sections of countries the economy of which lacked diversification. In some sections of the United States abandoned mines, depleted soils, and cutover forests have left ghost towns or stranded people dependent on relief from other areas. For whole countries these conditions may result if resources are depleted by foreign capital without sufficient interest in the lasting development of native economies. Such problems exist for the depletion of oil in the Near East; of copper, tin, and nitrate in some South American countries; and for exploitative forestry and plantation agriculture in colonial areas. This relation between resource use and economic instability will be discussed when the international aspects of conservation policy are considered in chapter 20.

Finally, resource use has been regarded as a cause for economic instability by numerous students who believe that variations of harvests in agriculture or other cyclical surpluses or shortages of primary raw materials are wholly or partly responsible for economic fluctuations—both for shorter "business cycles" and for longer "periods of expansion and stagnation." [1] Some point can be made for such an analysis within certain regions before the industrial revolution and before the great changes in transportation facilities during the eighteenth century tended to equalize supplies. At that time good harvests and prosperity and poor harvests and depressions tended to be associated. For the last hundred years, however, it can be shown that pros-

---

[1] Periods of expansion and stagnation, as the terms were used by the present writer, denoted temporary "cyclical" phenomena in economic development. See S. V. Ciriacy-Wantrup, "Agrarkrisen und Stockungsspannen zur Frage der langen 'Welle' in der wirtschaftlichen Entwicklung" [Germany] Reichs- und Pr. Ministerium für Ernährung und Landwirtschaft, *Berichte über Landwirtschaft*, N.F. 122. Sonderheft (Berlin: Paul Parey, 1936), 445 pp. Also: *Agricultural Depressions and Periods of Industrial Stagnation*, translated by C. O. Hardy. (Washington: Brookings Institution, 1938), 296 pp. Processed.

See also:

S. V. Ciriacy-Wantrup, "Major Economic Forces Affecting Agriculture with Particular Reference to California," *Hilgardia*, Vol. 18, No. 1, December 1947, pp. 1–76.

S. V. Ciriacy-Wantrup, "Resource Conservation and Economic Stability," *Quarterly Journal of Economics*, Vol. 60, No. 3, May 1946, pp. 412–452.

Later, in American and English economic literature, the term "stagnation" was used for a specific secular change of data—the thesis of economic maturity.

perity and depression in agriculture are more often the result than the cause of economic fluctuations in nonagricultural industries.

In this chapter another aspect of the relations between resource use and economic instability will be considered, namely the effects of economic instability upon conservation and depletion. Economic instability is one of the forces which influence conservation decisions (conservation and depletion *ex ante*) and which has important unplanned effects upon the time distribution of resource use (conservation and depletion *ex post*). To be sure, economic instability and resource use are interrelated variables either of which may be the dependent one for the purpose of analysis. That we take resource use as the dependent here does not mean the reverse aspect of the relation is unimportant for economic analysis. The focus on instability as an economic force resulting in conservation or depletion is merely suggested by the general frame of this part of our study (Part III) and by the fact that little attention has been given to it in the literature.

The Meaning of Economic Instability

What is meant by economic instability? A limitation of its many meanings is necessary for our purpose.

Individual firms and industries are born, grow, decay, and die because of changes in technology, in tastes of consumers, in institutional conditions, in personal leadership, and in other specific factors. No general statement about the effects of such changes upon conservation and depletion can be made; each individual change needs specific analysis.

During their lifetime, firms and industries are subject to instabilities which are an unavoidable part of productive processes. Such instabilities are variations of harvests because of weather conditions, bearing cycles of tree crops, and breeding cycles in wildlife and domesticated animals. These instabilities have general effects upon conservation because they tend to increase uncertainty allowance in production plans. However, they are temporary deviations from a secular norm that can serve as a basis for conservation decisions. Moreover the effects of production cycles on income are frequently mitigated by price changes in the opposite direction.

There are other instabilities for which such a norm and such a mitigating force do not exist. These instabilities are called here "economic fluctuations." Without implying strict periodicity, such fluctuations comprise both those of relatively short duration —so-called "business cycles"—and the longer ups and downs consisting of periods of expansion and stagnation. Economic fluctuations affect producers of primary raw materials in agriculture, forestry, mining, and the like, as violent changes of demand—that is, as violent changes of product prices which cannot be explained by changes in production and supply. "Economic instability," as the term is used in this chapter, refers mainly to these fluctuations. Other forms of economic instability will be considered in connection with international aspects of conservation policy.

Four Major Effects upon the State of Conservation

An analysis of economic fluctuations *per se* is outside of our present interest. We are interested here in how the state of conservation is affected. Four major effects may be differentiated.

First, economic instability increases the uncertainty allowance for most basic data used in production planning. An increase in uncertainty allowance generally leads to resource depletion. The reasons were explained in chapter 8.

Second, economic instability tends to raise the level of interest rates in the markets for loanable funds that are actually accessible to resource users. Uncertainty allowance of savers and lenders and imperfections in the markets for loans are generally increased. A higher level of interest rates tends to induce depletion.

Third, the decrease of incomes that takes place during the downward phase of economic fluctuations tends to increase time-preference rates of resource users. Why such an increase leads to depletion under imperfect markets for assets, and why this effect is not offset by conservation during prosperity, was explained in chapter 7.

Fourth, depletion of resources may occur if a depression forces contraction of production. In resource utilization such contraction is frequently not a smooth and quick adjustment to decreased prices of products, but a long-drawn-out struggle. Contraction is brought about not so much by the effects of decreases in product prices upon marginal revenues in production planning (*ex ante* effects), but by depletion of natural resources, equipment, and human fiber (*ex post* effects).

Difficulties of contraction in resource utilization are largely connected with aspects of the cost structure. Although some of these aspects were touched upon earlier, it is necessary to spell out in some detail the role of costs in the adjustment of resource use to economic instability. This will be done first for agriculture and then for mining. Such a separate treatment is advisable because differences exist between flow and stock resources and also between industries where the labor of the planning agent's family forms a large part of the total labor input and industries in which such labor is negligible.

Adjustment to Economic Instability in Agriculture

The failure of agricultural production to respond significantly and quickly to decreases of product prices (sometimes the response may even be an expansion of production) is discussed in economic literature as an inelastic, or, if a decrease of product prices elicits an increase in production, a negative supply curve. Problems of magnitude and size of (short-run and long-run) supply elasticity should not be confused with problems of lags and shifts in supply in response to price changes. A supply curve (short-run and long-run) is a concept of instantaneous (simultaneous) economics. The response of agricultural production to price changes is a continuous process extending over time. Strictly speaking, neither short-run nor long-run assumptions can be applied to this process. A negative supply curve actually consists of downward shifts of the supply curve in response to decreases of product prices. The reasons for such a situation will be explained presently. As a first step, we may look at the causes for an inelastic instantaneous supply curve of agricultural production as a whole, and of individual crops if they are produced as monocultures.[2] Clarification of these causes will throw light upon the problems of lags and shifts in supply. To the latter problem we will proceed as a second step in our analysis.

Adjustment in Instantaneous Economics

The importance of imperfections in the markets for farm loans and for farm assets has been emphasized earlier. The effect of such imperfections upon the adjustment of supply (in cases where adjustment requires outside capital or changes of assets) is fairly obvious. This effect is not peculiar to agriculture. To focus on the behavior of costs appears more profitable. Supply curves (short-run and long-run) can always be interpreted as (individual or aggregate) marginal cost curves. There are several aspects of the cost structure in agriculture which are relevant here.

In the first place, fixed and lumpy costs are particularly important in agriculture. Hired labor is sometimes a fixed item because of institutional conditions (as in parts of Europe), or because of employment contracts (as on rubber and sugar-cane plantations in colonial areas). Under certain assumptions, the labor of the planning agent's family may also be regarded as a fixed item. Usually, however, it is more appropriate to regard labor (both hired and family) as a variable cost of a kind to be discussed below. If fixed and lumpy costs are important in relation to variable costs, the point of minimum average variable costs (that is, the point at which production would cease in an individual enterprise, if product prices decreased further) may be far below the point of minimum average total costs. The fixed-cost argument for maintenance of supply under decreasing product prices is valid only under short-run assumptions.

Another aspect is connected with the slope of the marginal cost

---

[2] The price elasticity of agricultural production as a whole is especially small, because of various effects of joint costs. These effects were considered elsewhere.—S. V. Ciriacy-Wantrup, "Economics of Joint Costs in Agriculture," *Journal of Farm Economics*, Vol. 23, No. 4, November 1941, pp. 771–818.

On the other hand, individual crops grown in a diversified system of farming may show a considerable price elasticity of supply, because shifts between crops and livestock are relatively easy in such a system.

curve. The effect of price changes upon production between the points of minimum average total costs and minimum average variable costs depends upon the slope of the marginal cost curve. In agriculture the slope between these two points is frequently steeper than in manufacturing. Tendencies toward diminishing marginal physical output are generally more powerful in agriculture.[3] Changes of variable inputs in response to changes of product prices, therefore, may affect production but slightly. This aspect of cost structure is generally more important under intensive than under extensive variations of supply;[4] from our standpoint extensive variations are more relevant. The argument of diminishing returns can be used under long-run as well as short-run assumptions.

Furthermore, in agriculture, important variable costs are noncash inputs, in the sense that they need not be purchased at the market, nor is their market value readily ascertainable. These variable costs are those of the operator's services, of family labor, of feed, and of soil fertility. Markets for the services of the operator and his family are highly imperfect, because of specialization and indivisibility in these services, and because their alternative opportunities are closely related to the likewise imperfect markets for farm assets and loans—especially during depressions. Feed can be produced on the farm. Fertilizer purchases can be avoided by relying on depletion of soil fertility. The noncash character of these variable costs (and of corresponding returns—for example, food and shelter, which occur jointly with cash returns) is important for agriculture's response to price changes, because inequality of marginal inputs and outputs is difficult to detect. When prices decline, such inequality is neither apparent

---

[3] The reasons for this "law of diminishing returns" are technological. They are largely connected with the role of space (both in its horizontal and vertical dimensions) and of biological factors. Both aspects are more important in agriculture. In manufacturing, some studies indicate rather small slopes of marginal cost curves. See Committee on Price Determination, *Cost Behavior and Price Policy* (New York: National Bureau of Economic Research, 1943), 356 pp.

[4] "Extensive" refers to variations of supply brought about by increases in the acreage of individual crops, either at the expense of other cultures or by bringing new land into use. "Intensive" refers to variations of supply without changes in acreage of individual crops.

nor dangerous through its effects upon the financial liquidity of the enterprise. This factor also operates under long-run as well as short-run assumptions.

Finally, in agriculture, productive services which give rise to fixed, lumpy, and noncash variable costs frequently constitute a complete production unit. Minimum average variable cash costs may be close to zero to the right of origin in a Cartesian system of coördinates. In other words, production will continue if price declines of products reduce or even prevent purchases of hired labor, feed, and fertilizer.

These factors, together with factors of jointness, explain the rationale of small instantaneous (short-run and long-run) supply elasticities—especially if prices decline. They do not explain so-called negative supply curves. Moreover, as already indicated, we are interested here especially in lags of production adjustment and in shifts of supply over time. Let us now consider these two problems in turn.

Adjustment in Time Economics

If the time dimension of fixity is considered, fixed costs and lumpy costs may both be discussed as sunk costs. This term has the further advantage of including both noncash and cash variable costs.[5] In agriculture, variable costs are sunk over various periods of gestation. A high proportion of sunk costs affects production adjustment over time in the same way that a high proportion of fixed costs affects supply elasticity: sunk costs are likewise disregarded in production adjustment. Additional variable costs, which in this case may be called "recovery costs," are expended until their marginal unit equals the marginal unit of corresponding revenues. We might repeat the argument offered above in the case of fixed cost, substituting the terms "sunk costs" for "fixed costs" and "recovery costs" for "variable costs."

Shifts of supply curves over time in periods of decreasing

[5] As will be recalled, the terms "fixed," "lumpy," and "variable" refer to instantaneous variations in rates of production, not to time (chap. 5). However, the dimension of time (use over more than one interval) is implicit in the differentiation between fixed and lumpy costs on the one hand, and variable costs on the other.

prices of agricultural products are explained largely by factors that affect hired and family labor input. Prices of other inputs (fertilizer and equipment) have generally been less flexible than product prices. There is some evidence that technological improvements have spread faster during depressions than during prosperity; but the evidence on this point is not conclusive.

During economic fluctuations, in most countries, wage rates for hired labor have been more flexible in agriculture than in other industries. In industrial countries (for example, the United States and western Europe) this fact is related to less rigid unionization of the agricultural labor market, to the greater importance of payment in kind (room, board, agricultural products, land to produce food and feed) and, partly induced by these differences, to rural-urban and urban-rural migration during economic fluctuations. In colonial areas operating with imported labor (for example, Chinese and Indian labor in Burma, Malaya, and the Dutch colonies), wage rates were rather flexible during depressions. This occurred because alternative local employment was difficult to obtain, and because remigration was frequently not desired by the laborers themselves.

Family labor input, including labor of the planning agent, is determined not by the movement of wage rates but, in view of the market imperfections for farm assets and loans already mentioned, by the balance of marginal disutility of effort and marginal utility of product. This balance may be decisively affected if decreases of product prices increase the marginal utility of money. As we know, this effect becomes more and more important as incomes decline. Family labor input, therefore, may increase in response to decreasing prices, especially in the low-income groups.

This shift of the supply curve for family labor during agricultural depressions is not identical with the increase of time preference mentioned as the third effect of economic instability upon the state of conservation; but it has the same causes and similar results. Both phenomena are related to variations in the marginal utility of money in the course of income declines. An increase of time preferences, however, may have the above (third, p. 202) effect upon the state of conservation, even if the planning agent and his family do not furnish a significant part of the total labor input.

Consequences for Soil Conservation

To summarize: most characteristics of the cost structure tend to make agriculture's response to price changes sluggish as compared with manufacturing. This influence is much more pronounced when prices decline than when they rise. Under certain conditions, a response to declining prices may be absent for a long time or may be negative, especially among low-income, self-employed farmers. Output is maintained or even increased by depleting the soil and human fiber. A vicious circle—low prices, low incomes, soil depletion—is created.

This situation is especially serious for the soil (and for the farmer) if, in response to a period of high prices, production has expanded into areas where permanent cultivation is exposed to particularly severe climatic or topographic soil-erosion hazards; it is also serious if monocultures have become dominant in regions where soil depletion can be avoided only through a diversified system of farming.[6] The more spectacular examples were caused by war-time inflation of product prices.

The First World War caused a rapid expansion of wheat acreages in the semiarid portions of the High Plains where wind erosion became a serious problem, of bean and pea acreage in the Coast Range of California where topographic erosion hazards are high, of sugar-cane acreage in Cuba and Puerto Rico, and of cotton and tobacco acreage in the Piedmont of southeastern United States—that is, in subtropical areas where climate, soil, and topography combine to create erosion hazards. The Second World War and its aftermaths caused an expansion of wheat and soybean acreage which in the opinion of some students will create problems of soil depletion similar to the erosion in the wake of the First World War.

If declining prices caused a quick contraction—that is, a shift

[6] Diversification of farming systems should not be confused with diversification by regions. From the standpoint of soil conservation the former is usually more relevant.

to more extensive types of land use that are better suited to the physical conditions—such as permanent grazing, production of forests and wildlife, a more diversified agricultural economy with emphasis on legumes and livestock—depletion would not become serious. The real cause for such depletion is the difficulty of a quick and smooth adjustment of agricultural production to price declines in the course of economic fluctuations. This difficulty is connected with the behavior of costs.

Adjustment to Economic Instability in Mining

Turning now from agriculture to mining or, more generally, from flow to stock resources, it should be noted that economic instability affects the state of conservation in certain ways which, like the effect just observed in agriculture, are closely connected with difficulties of contracting production in orderly fashion—that is, from the standpoint of this study, without wasteful depletion of resources. As in the case of agriculture, these difficulties are connected with peculiarities in the cost structure.

In the mineral industries, mines or sections of mines that become temporarily submarginal during a depression must be properly maintained if operation is to be resumed later. Water intrusion (potash, bauxite, zinc), cave-in (coal), seepage and loss of pressure through blowoff (oil, gas), and rapid deterioration of unused fixed and specialized equipment (shafts, wells, dredges, transportation facilities, pumping and air-conditioning systems, ore-improvement plants) are the main factors that prohibit a resumption of operations without proper maintenance. Alternatively, these same factors necessitate rather high costs for maintenance. If restriction of output is brought about, not by planned adjustment to decreases in product prices but forcibly through inability of high-cost producers (high in terms of minimum average cash costs and of fixed charges) [7] to meet current obligations after exhaustion of financial reserves, costs of main-

[7] Fixed charges represent contractual obligations of an economic unit to make current payments regardless of the nature and origin of these payments. Fixed costs, on the other hand, can be attributed to the manner of use of productive services regardless of property and other rights (chap. 7).

tenance cannot be met. A decrease of product prices, therefore, may lead to permanent destruction of resources which would not be submarginal after the depression has passed. The cost structure of mining makes such wasteful death struggle of temporarily submarginal capacity almost certain under *laissez faire.*

In mining, fixed, lumpy, and sunk variable costs constitute a major portion of the total. Interest and amortization charges for fixed and specialized equipment loom large. More importantly, fixed costs of maintenance—that is, costs for maintenance at zero production—often increase more than proportionately with sunk investment. The reason has already been implied: The more deposits of stock resources are opened up through investment in shafts, wells, and removal of protective strata, the greater the danger from water intrusion, seepage, blowoff, and cave-in.[8] The importance of these aspects of the cost structure for smallness of supply elasticity and for lags of supply need not be repeated.

In mining, variable costs consist largely of labor costs. As in agriculture, the increase of marginal costs is rather steep over the relevant range, because of the "law of diminishing returns." The technological reason for decreasing marginal physical output in mining is largely connected with the role of space (in its vertical rather than its horizontal dimension) and of grade of ore.[9]

Noncash variable costs and shifts of supply curves during depressions are less important in mining than in agriculture. Input of family labor in relation to total labor input is negligible. Hired labor is well organized and paid in cash—as in manufacturing. An exception is found in mining enterprises in colonial areas operating with coolie labor.[10]

This exception, although generally not sufficient to cause negative responses of production to price, leads to another point that

---

[8] For example, the Alsace-Lorraine potash deposits and a United States bauxite deposit discussed in chapter 3.

[9] For illustrations see Andrew V. Corry and O. E. Kiessling, "Grade of Ore," *Mineral Technology and Output per Man Studies* (Philadelphia: Works Project Administration, National Research Project. Report No. E-6, 1938), 114 pp.

[10] In some areas, for example, in Malayan tin mining before the Second World War, labor conditions of Indian coolies were closely supervised by the government. Rather than approve wage decreases, the government sometimes forced the operators to repatriate the laborers.

is relevant here. In mining, great differences in cost structure exist between individual enterprises of the same industry. More important than differences in labor organization are differences in technological conditions (grade of ore, depth, width, and total quantity of deposits, geographical location, and accessibility of mines) and in financial strength. Because of these differences, individual enterprises vary greatly with respect to profitability and the financial necessity of curtailing production. There is a tendency to delay a decision about shut-down until financial reserves are exhausted or appreciably diminished. Competition is frequently imperfect. During a depression, producers are reluctant to relinquish their share of the market to others. They fear that such a loss may decrease their chances of participating in the recovery. Generally, duration and severity [11] of decreases in product prices are underestimated in producers' expectations. An attempt is made to hold on a little longer, in the hope that the turn of events is near.

For some stock resources, especially metal ores, a considerable and increasing part of the total current supply of the refined product originates from scrap. The supply of scrap is largely determined by the consumption of virgin resources in the past and by the rather erratic rate of obsolescence of finished goods. Current price elasticity of scrap supply, therefore, is rather small over the relevant range. The burden of adjusting total supply (from primary and secondary sources) falls mainly upon the primary producers.

The conditions just described have several consequences. Enterprises which are forced to shut down after a long death struggle are usually not in a position to meet expenditures for proper maintenance. As already explained, this may lead to the destruction of resources. These resources may be ones that would not have been submarginal after the depression has passed, and enterprises eliminated may not be high-cost producers. The weeding-out process may reduce future supply much more than prospective demand warrants and, in terms of social contributions, it is not

[11] During depressions, demand curves for raw materials not only shift downward but also decrease in elasticity.

necessarily the fittest who survives. Survival may be determined by accumulations in the past, by financial connection with other industries, and by exploiting a monopsonistic position in the labor market. Thus, although the industry may have permanently submarginal capacity which should be eliminated, to have this accomplished through a struggle for survival may not be at all desirable from the standpoint of conservation.

Possibilities for an Anticyclical Conservation Policy

It may now be asked how far is it possible to modify the depletion of resources in the course of economic fluctuations by measures of conservation policy. The answer to this question may be given in two parts. First, we will consider whether and how conservation policy can contribute to greater economic stability. Second, we will consider what can be done about the four effects of economic instability enumerated earlier, assuming that anticyclical policies will not reduce economic instability sufficiently to eliminate its effects on conservation.

From what has been said it follows that public policies which attack economic fluctuations at their roots—that is, bring about greater stability in the demand for the products of primary industries—are of great interest for conservation. In recent decades considerable progress has been made in analyzing the two types of economic instability which are relevant for conservation. This analysis employs the familiar set of tools which were forged in their basic shape by Knut Wicksell [12] and later sharpened by American, Swedish, German, and English economists. This set effectively combines the overproduction, underconsumption, and monetary theories of economic instability in a theory of saving, investment, and income formation.

Public policies which aim directly at the balance of saving and investment go far beyond the field of conservation and cannot be discussed here. However, conservation may furnish an important outlet for investment. Thus, if conservation policy is properly integrated with more general policies that affect the volume of saving and investment, greater economic stability may

[12] Knut Wicksell, *Geldzins und Güterpreise* (Jena: Gustav Fischer, 1898), 189 pp.

be obtained as an important joint product. What are the requirements of an anticyclical investment policy and what are the possibilities of fulfilling these requirements through investment for conservation?

Timing, Volume, and Flexibility of Investments

Judging from experience as well as from economic theory, the decrease in employment, income, and prices that results from failure to offset savings is cumulative up to a certain point in its development. An anticyclical investment policy must, therefore, have accurate timing, sufficient immediate volume, and great flexibility in view of rapid changes in the demand (from private planning agents) for savings.

Satisfaction of the requirement for proper timing, volume, and flexibility depends largely upon the technical and economic possibility of telescoping [13]—of concentrating an anticipated long-range program within a shorter period. Conservation is superior to many other public works in this respect. Future needs for conserving most natural resources can be filled in advance without fear that the result will be diminished over time through nonuse. With many natural resources, the results of conservation efforts appreciate rather than depreciate during nonuse. This is especially true of soils, forests, and grasslands. In contrast, future needs for housing, for public utilities, for roads, cannot be satisfied in advance for very long, even if they are correctly anticipated. Without use, deterioration or cost of maintenance or both would be too high.

Another aspect that is important in regard to timing, volume, and flexibility is the possibility of sectionalizing. By this term is meant the division of investment into economical units. Each section should be of a kind that can be completed at any time when completion appears desirable from the standpoint of anticyclical public policy. From this standpoint, again, conservation of soils, forests, grasslands, wildlife, and minerals, is well suited.

[13] This term is used in Benjamin Higgins, "Problems of Planning Public Work," in S. E. Harris, ed., *Postwar Economic Problems* (New York and London: McGraw-Hill Book Co., Inc., 1943), pp. 187–205.

Such works can be undertaken in sections without detracting from the value of the project as a whole. On the other hand, a half-finished housing project, or an incomplete system of public utilities and roads, or a partly constructed dam is not practical.

Coördination between Private and Public Investments

Next to the requirements with respect to timing, volume, and flexibility, we may consider the necessity of proper coördination between public and private investment. Since it is the aggregate of public *and* private investment that counts, public investment should choose outlets of such a kind that private investment will be encouraged rather than discouraged. Such coördination requires investment outlets in which public investment is complementary to, but does not compete with, private investment. Public investment should be concentrated, first, on the production of goods and services that are demanded and supplied collectively (they are largely extramarket goods; chapter 17); second, on the industries that are under public control.

In these respects conservation of resources is also superior to other fields of public investment, because it is largely neutral or complementary to private investment. Extramarket values in conservation are especially important. Some socially highly significant resources, such as forests, grasslands, wildlife, and water, are partly in public ownership or under direct public control.

Labor-intensive and Material-intensive Investments

The third general requirement of public investment is proper use of labor-intensive and material-intensive investments. These terms mean that some kinds of investment lead to a large direct increase in employment, whereas other kinds require more raw materials, construction, equipment, and machinery, and thereby lead only indirectly, usually with some lag, to a large increase in employment. Sometimes, in economic fluctuations, public investment of the labor-intensive type is called for. This is true, for instance, when abnormally large frictional unemployment exists during the reconversion of industry from war to peace. In such a situation, material-intensive types of public investment would compete with private industry for scarce raw materials, equipment, and tools, would accentuate bottlenecks, and would increase the danger of inflation. In another situation—for example, during the downward phase of the business cycle or in a stagnation period—public investment of the material-intensive type would be more desirable, because of its stimulating effects upon private investment.

In conservation, some investments are labor-intensive and others material-intensive. Hence with proper selection of projects, conservation expenditure is well suited for different phases of economic fluctuations. Soil and forest conservation, for example, by a Civilian Conservation Corps using already available heavy equipment (war-"surplus" tractors, graders, and other earth-moving machinery), is labor-intensive. Measures of water conservation, irrigation, and flood control (water reservoirs, check-dams, stream-channel rectification, hydroelectric developments) are examples of material-intensive investments.

Regional Distribution of Investments

The fourth general requirement of public investment policy is proper regional distribution. Unemployed labor resources are not freely movable. They may be temporarily unemployed in a "permanently" superior location. On the other hand, permanently superior alternative employment may require a change of location, and public investment may be utilized to stimulate such a movement.

Public investment in conservation cannot be regionally directed at will. Such investment is always location-bound. In this respect, therefore, conservation outlets are inferior to other outlets of public investment. Regional developments—for example, by the widely discussed river authorities—are long-range programs to relieve depressed areas or to open new frontiers for investment in a secular perspective. But such developments cannot easily be fitted into anticyclical investment policies to alleviate temporary unemployment in specific localities.

Investments and the Balance of Payments

Finally, the fifth requirement of a successful public investment policy is avoiding adverse repercussions upon the balance-of-payments position of the country engaged in such a policy and upon international monetary stability. In this respect conservation of United States stock resources deserves a preferential position as an investment outlet. Such conservation requires long-term capital exports for resource development abroad, removal of direct and indirect export subsidies on domestic resources, and lowering or, better still, abolishment of tariff barriers on imports of foreign stock resources. Such policies appear desirable from the standpoint of the balance-of-payment position of the United States and its trade partners, and from that of monetary stability throughout the world.

Mitigating the Effects of Economic Instability

Anticyclical policies are in the formative state. Even after they have been more fully developed, it cannot be expected that economic instability will be reduced sufficiently to be of no concern for conservation. The question arises, therefore, how can the four effects of economic instability which were mentioned above be mitigated?

Possibilities for public policies to mitigate the first three effects of economic instability upon conservation—through uncertainty, interest, and time preference—were discussed earlier. Rather than repeat in a somewhat different connection what is already familiar, a few words may be said about the fourth effect of economic instability upon conservation—through difficulties of contracting production quickly and smoothly. In this case the objectives of conservation policy are control of shortsighted expansion and assistance to orderly contraction. These objectives are frequently claimed for domestic and international policies.

Domestic Policies

Domestically, in the United States, there are a few encouraging beginnings. As examples may be mentioned the increased use of zoning ordinances; the techniques (rather than the objectives) developed in agricultural production controls; public assistance to agricultural adjustment through education and the program of purchasing submarginal land for use as public forests, game refuges, and watersheds;[14] the withdrawal of public domain through the Taylor Grazing Act; and the control of oil and gas production through state laws and state compacts. These and other domestic tools of conservation policy will be considered more fully in chapter 19.

Public policies in the United States have generally been more concerned with protecting resource users against price decreases than with controlling shortsighted expansion during periods of increasing prices and with facilitating adjustment during periods of decreasing prices. Relief payments, parity prices, production subsidies, and credit assistance for rehabilitation were designed primarily to protect existing patterns of resource use and not to assist in their change.

International Policies

In the international field the situation is similar. Encouraging beginnings are the international conventions concerned with the conservation of fugitive resources (fisheries, migratory birds, and upland game). Although they are not mainly concerned with the effects of economic instability, they are well adapted for obtaining some objectives which are relevant here: Such conventions could be used to control shortsighted expansion of production (for example, through depletion of breeding stock in fisheries) during periods of high prices. During depressions they may facilitate the maintenance of those conservation practices which decrease current income in favor of future income (for example, regulations regarding size of nets, closed seasons, use of explosives and of stationary traps). Some of these conventions have already proved their effectiveness for obtaining these objectives.

[14] A modest start was made under the Weeks Law of 1911. A new impetus was provided by funds appropriated under the National Recovery Act of 1933, the Emergency Relief Act of 1935, and especially under Title III of the Bankhead-Jones Act of 1937, mentioned earlier (chap. 11).

These and other international tools of conservation policy will be considered more fully later.

International commodity agreements in the raw-material industries have similar *potential* value for conservation policy: During booms, they could help to restrict expansion of production in accordance with physical conditions (for example, of cultivated acreage in areas with climatic or topographic erosion hazards). During depressions they could help in orderly contraction (for example, by putting temporarily submarginal mining capacity on a stand-by basis and providing for proper maintenance). In general, they could facilitate international adoption of conservation practices (for example, silvicultural cutting practices as recommended in the NRA Code, drilling and pumping practices as stipulated in some state oil and gas conservation codes).

Although the word "stability" is very popular as a stated objective with all commodity agreements, it will be shown below that the historical evidence speaks against their effectiveness in this respect. It appears highly doubtful whether sufficient price stability (that is, from our standpoint, sufficient in degree and duration to affect the state of conservation) can be obtained by adjustments of supply, without attacking the causes of economic fluctuations at the roots. To counteract violent changes of demand, which are at the roots of economic fluctuations in raw-material industries, through planned changes of supply, requires almost superhuman knowledge, foresight, integrity, and administrative efficiency of planners, and a rare comprehensiveness and thoroughness of organization in order to prevent production by outsiders and noncompliance of members. Required, too, is a great flexibility of supply. Usually, such flexibility requires large [15] buffer stocks, because of familiar factors in the cost structure.

For all these reasons, there appears little merit in arguing here the point whether or not commodity agreements are potentially capable of increasing stability of prices through adjustments of

[15] Proponents of a centrally controlled buffer stock often overlook the fact that it will partly replace stocks held by the trade and by major consumers. These groups are only too glad to shift the risks and costs of carrying stocks to producer organizations or to the public. In order to have the desired additional effect, therefore, a buffer stock must be large.

supply. It is more relevant to consider the relations between commodity agreements and conservation under the assumption that price fluctuations cannot be sufficiently (in the defined sense) reduced by such agreements. This does not mean that commodity agreements can have no effects upon conservation through increasing (or decreasing) stability of product prices. It does mean that the conservation argument for commodity agreements must be built on a more realistic foundation. As already indicated, commodity agreements have potential value for conservation policy quite aside from their effects upon price stability.

PART IV

THE SOCIAL ECONOMICS OF CONSERVATION

OBJECTIVES AND CRITERIA OF CONSERVATION POLICY

CHAPTER

16

CONSERVATION POLICY AND SOCIAL INSTITUTIONS

Definition of Conservation Policy

"Policy" has almost as many different meanings as "conservation." It seems helpful for our purposes to restrict the term "policy" in accordance with its etymological origin as relating only to actions by the organized public, whether nation, state, county, city, or public district. Any private individual, a firm, association, parties may have opinions, attitudes, proposals pertaining to policy; they may aid in forming and executing such a policy; and they are always affected by it; but the term itself will be used here in the restricted sense of *public* policy.

Thus, "conservation policy" as a field of scientific inquiry may be defined as a study of actions of governments at different levels with reference to the intertemporal distribution of use rates of resources. The study is concerned with the formulation, the execution, the criteria, the objectives, the implementation (tools), and the effects of conservation policy.

Formulation and Execution of Conservation Policy

In a democracy, conservation policy is formulated—although not necessarily originated—by the legislative and carried out by the executive branch of governments at various levels.[1] If the branches of government are not formally distinguished, as in authoritarian states, the formulation and execution of policies are nevertheless separated in planning and action departments.

How the policy-formulating body of the government is constituted—for instance, through popular election of representatives in its numerous variations, through appointment by an autocrat, or through direct representation of all members of a social class (nobility, priesthood)—is a question of social weighting. This problem in its relation to conservation policy will be discussed in chapter 17.

Formulation and execution of conservation policy, although separate in theory, are interrelated in practice. To use Maine's words, ". . . substantive law has at first the look of being gradually secreted in the interstices of procedure . . ."[2] Execution of a policy, the mandate for which is received by the executive from the legislature, usually requires much interpretation and formulation in details. Furthermore, formulation of a policy is influenced by the experience gained in its execution. The executive may take the initiative in bringing a certain policy before the legislature, not only in authoritarian states, but in democracies like Great Britain. On the other hand, the legislature may pre-

---

[1] In the United States the judiciary also plays a role in formulating, limiting, and redirecting policies. This role stems from the dominant influence of the common-law method of thinking—symbolized by precedent, custom, special case, and concrete assumptions—upon the approach to social problems in general. It is in contrast to the European legal system, which is dominated by codes (Roman Law, Code Napoléon) and by abstract assumptions that are changeable only by the legislature and not molded by the judiciary. In European countries and in Latin America, which are strongly influenced by Roman Law and the Code Napoléon, the judiciary has less influence on policies.

[2] Sir Henry James Sumner Maine, *Early Law and Custom* (London: John Murray, 1883), p. 389.

scribe not only the policy, but also the tools of its execution, and may supervise it through special agencies of its own (congressional committees, public hearings).

Formulation and execution of a policy are usually influenced by those who are positively or negatively affected by it—that is, by pressure groups. Influence is exerted not only through representatives in parliaments, but also through occupational and other organizations and even individuals in direct contact with departments of the executive, and through the somewhat elusive but highly important channels of public opinion. Even policies formulated by an autocrat and his appointed legislative council are not exempt from these influences. On the other hand, the government may use public opinion to educate those who are affected by its policies. Such education stretches from occasional press conferences to the far-reaching activities of a ministry of propaganda.

In short, social institutions determine strongly the formulation and execution of conservation policy. Yet, in the preceding chapters, social institutions and their economic effects were mentioned as obstacles to be overcome—that is, as objects of conservation policy—or as tools to be employed by that policy. The problem arises, therefore, whether the latter point of view is justified: whether and why social institutions should be regarded as variables rather than constants for purposes of conservation policy. In this study the pragmatic position is taken that both points of view are possible: in a study of human ecology, determinism and free will may both be helpful as working hypotheses.

Differences between a Social Institution and a Policy

The two basic aspects of a social institution, its "concept" and its "structure," were discussed earlier. A similar differentiation applies to public policy if the terms "concept" and "structure" are translated into "objectives" and "tools"—the terms used in this study. This parallelism of terms carries the danger that a "policy" may be confused with an "institution." In fact, a prominent representative of the American institutional school of economists defines an "institution" simply as "collective action." [3] This would make the meaning of "institution" identical with that of "policy." Thus, it is necessary to note some differences between them. This must be done without recourse to the hard-and-fast definitions that are so difficult in view of the great variety of social institutions.

An existing policy implies objectives of social action for the present and the future. Objectives—in Sumner's terms, the "concept"—of existing social institutions may relate only to the past and may be compromised or lost in the course of time. But the "structure" may still keep the institutions effective. Institutions represent the residual agglomerations of *past* social actions and experience gained from these actions. A policy implies objectives of social action for the *future.* (Such objectives—and attempts to realize them—may, of course, be studied as historical phenomena.)

Closely related to this first difference is another. The term "policy" as defined above implies that important groups of individuals perceive its meaning with respect to both objectives and tools. Individuals may perceive the concept and structure of institutions only if the institutions are in conflict with present interests: "For such casual glimpses of the intricacies of social institutions as men are permitted to see they are indebted to the stress and strain of transition." [4] This difference in consciousness usually means that individuals make fewer deliberate choices under the influence of social institutions than under the influence of policies. True, their behavior is determined by the social compulsion inherent both in institutions and in policies. But in social institutions "automatic submission" [5] or what we have called "habit patterns" are generally more important in influencing individual behavior. In policies, the individual usually chooses submission

[3] ". . . we may define an institution as collective action in control, liberation and expansion of individual action."—John R. Commons, "Institutional Economics," *The American Economic Review,* Vol. 21, No. 4, December 1931, p. 649.

Cf. also Common's comprehensive text *Institutional Economics* (New York: The Macmillan Co., 1934), 921 pp.

[4] Walton H. Hamilton, "Institution," in *Encyclopedia of the Social Sciences,* Vol. 8, p. 88.

[5] Cf. Joyce O. Hertzler, *Social Institutions* (New York: McGraw-Hill Book Co., 1929), pp. 142 ff.

after consciously considering alternatives that to him appear less desirable.

The aspect of social compulsion directs attention upon still another difference between policies and institutions. Policies are invested with the sanction and authority of the governments which, at various levels, formulate and execute them. Many institutions, however, do not concern governments at all. Compliance is secured by loyalty, by self-interest, and by public opinion—by the fear of appearing ridiculous, clumsy, or merely a poor sport.

Finally, differentiation between economic and other policies (educational, religious, military, family policies) can be drawn fairly well. Economic institutions, on the other hand, are intricately interwoven with social institutions in the other fields.[6] One group of scholars contends that economic interests are the basic factors in forming political, religious, educational, matrimonial, and other institutions. According to this view, all social institutions are essentially economic, formed and upheld by economic classes.[7] Another equally important group believes that religious and other noneconomic institutions determine economic ones.[8] Whatever these relations may be, the economic effects of social institutions are undoubtedly more widely diffused than the effects of economic policies.

[6] Hertzler (*ibid.*, pp. 47–64) differentiates between nine "fields" in which social institutions may be operative: (1) The economic and industrial field. (2) The matrimonial and domestic field. (3) The political field. (4) The religious field. (5) The ethical field. (6) The educational field. (7) The communicative field. (8) The aesthetic field. (9) The health field.

[7] Besides the writings of Karl Marx and his followers, a long list of non-Marxian works could be mentioned, of which the best known in the United States is probably Charles A. Beard, *An Economic Interpretation of the Constitution of the United States* (New York: The Macmillan Co., 1913), 330 pp.

[8] Max Weber, "The Evolution of the Capitalistic Spirit," in *General Economic History*, translated by Frank H. Knight (London: George Allen and Unwin, Ltd., 1927), pp. 252–269.

Max Weber, *The Protestant Ethic and the Spirit of Capitalism*, translated by Talcott Parsons (London: George Allen and Unwin, Ltd., 1930), 292 pp.

Werner Sombart, *Der moderne Kapitalismus.* (München und Leipzig: Duncker and Humblot, 1928), 3 vols.

Eduard Hahn, *Die Entstehung der Pflugkultur* (Heidelberg: Carl Winter's Universitätsbuchhandlung, 1911), 192 pp.

Social Institutions as Variables for Conservation Policy

From these differences we may draw some conclusions for the relation between conservation policy and social institutions. Even though formulation and execution of conservation policies are affected by social institutions, such policies can still aim at changing the influence of social institutions upon the state of conservation. As just indicated, that influence may no longer correspond to the interests of the social group. The original "concept" of the institution may be lost or compromised. The electorate may not even perceive the effect of social institutions upon the state of conservation, and the first step of conservation policy must be to make them perceive it.

Conservation policy, however, may also seek to alter those social institutions which still possess identity of "concept" and "structure," and which are clearly imprinted upon the consciousness of the people. This may happen if social institutions receive their authority from social groups other than governments. What is more important, economic change often produces conflicts between the widely diffused economic effects of social institutions. A lag of social institutions behind economic changes plays an important role in the study of history—Hegel's "historical dialectic." Institutional lag must produce, at some point in time, rapid and violent changes—social revolutions. From this perspective, conservation policy may attempt to bring the social institutions affecting the state of conservation into agreement with each other and with changing economic conditions. This aim may involve the weakening or elimination of some social institutions and the strengthening of others.

"Concept" and "structure" of those social institutions which have been discussed thus far—for example, those of property and of market form—are not primarily concerned with conservation. If such institutions are changed in the interest of conservation, the actual results (in terms of conservation and depletion) are uncertain in degree, geographical extent, and timing. Conservation policies which operate through social institutions not primarily concerned with conservation may be said to employ "indirect" tools.

There are other tools of policy which are primarily concerned with conservation. They "directly" induce planning agents to make conservation decisions: not only automatic submission or economic calculations but, in addition, the threat of police power and of court action operates as inducement. Such direct tools vary in type from mild forms of zoning to the requirement (or prohibition) of specific conservation (depletion) practices and the fixing of production and sales schedules. Before these tools of conservation policy are discussed more fully, the social objectives and criteria of conservation policy must be clarified.

CHAPTER 17

OBJECTIVES AND CRITERIA OF CONSERVATION POLICY

The Optimum State of Conservation in Social Economics

The objectives of conservation policy—that is, the optimum state of conservation in social economics—could be determined by criteria similar to those used for the private optimum, subject to similar limitations and approximations, provided revenues and costs could be reinterpreted for social accounting in a meaningful way. If this could be done, the objective of conservation policy could formally be defined as that state of conservation which maximizes social net revenues over time.

The problems encountered in making such a concept theoretically acceptable and practically workable are even more numerous and complex than in private economics. As in private economics, therefore, we will have to be content with approximations. However, in order to understand when and what kind of approximations must be used and in order to realize their limitations fully, the major problems of reinterpreting revenues and costs in social economics must be examined. The reader not particularly interested in the analytical difficulties encountered in "welfare economics" may want to pass over most of this chapter and continue with the approximations taken up from page 249 onward and in the following chapter.

Social Revenue and Cost Functions

In reinterpreting "revenues" and "costs" from the standpoint of social accounting, it is best to focus on marginal rather than total functions. The former have the most direct relations to commonly used functions in private economics. Furthermore, for our present problem—namely, maximizing social net revenues in resource utilization—the marginal functions are of particular relevance.

Marginal social revenue functions can be identified with community demand functions and marginal social cost functions with community supply functions. The term "community"—instead of "market"—is chosen because, as shown below, public policy is not necessarily confined to using functions which are directly yielded by the market.

The identification of marginal social revenue with demand functions and marginal social cost with supply functions is possible because certain value changes which are important in private economics are transfer items in social economics: Changes in prices of products, if functionally related to the rate of production (that is, not related to other influences, such as changes in the demand for products), constitute transfer items between producers and consumers of resources. No change in social revenues is involved. Likewise, price changes of productive services functionally related to the rate of production (that is, not related to other influences, such as changes in the supply of services) must be excluded from marginal social costs because they are transfer items between producers of resources and owners of productive services. No change in social costs is involved. If supply of a product is fixed (elasticity of supply being zero and remaining so over time), there are no marginal social costs.[1]

[1] If it is desired to make the identity of community supply function and marginal social cost function universal, prices charged to bring demand into line with the

With respect to the above interpretation of social costs, it may be well to emphasize that the concept of a social cost function implies that given production (use) rates are realized with minimum total social costs, subject to the technological conditions defined by the production function. This requirement exists in social economics no less than in private economics. Through this requirement, a cost function is always related to a production function.

Maximizing Social Net Revenues in Instantaneous Economics

In instantaneous (simultaneous) economics, the identity between marginal social revenues and demand and between marginal social costs and supply can be used for representing total social net revenues by the relevant area under the demand curve minus the relevant area under the supply curve. "Relevant" in this connection means the area to the left of the intersection between demand and supply curves. More than a hundred years ago Dupuit called these total social net revenues *"l'utilité."* [2] Fifty years later the same concept appears as Marshall's "consumer's and producer's surplus." [3] In this form the concept became discredited because of its alleged connection with the problem of measuring and comparing subjective utilities.[4] According to more recent views, Dupuit's approach to the maximization of total social net revenues is meaningful provided we make, with respect to the interrelation between different goods and money, the same assumptions that are generally implied in dealing with Cournot-

---

fixed supply may be termed marginal social costs. The latter terminology is employed, for example, by Harold Hotelling, "The General Welfare in Relation to Problems of Taxation and of Railway and Utility Rates," *Econometrica,* Vol. 6, No. 3, July 1938, p. 249.

[2] Jules Dupuit, "De la Mesure de l'Utilité des Travaux Publics," *Annales des Ponts et Chaussées, Mémoirs et Documents,* 2e Série, Vol. 8, No. 116, 2e Semestre, 1844, pp. 332–375.

Cf. also Jules Dupuit, *De l'Utilité et de sa Mesure; Écrits Choisis et Republies, par Mario de Bernardi* (Torino: La Riforma Sociale, 1933), 228 pp.

[3] Alfred Marshall, *Principles of Economics* (8th ed.; London: Macmillan and Co., Ltd., 1930), Book 3, chap. 6.

[4] A summary of this critical point of view is presented by Joan Robinson, *The Economics of Imperfect Competition* (London: Macmillan and Co., Ltd., 1933), pp. 211–217.

Marshallian demand and supply curves.[5] Other developments [6] of Dupuit's concept represent more a translation into the professional terminology of today.

Maximizing Social Net Revenues in Time Economics

In time economics, where joint production is always implied, a two-dimensional treatment of the maximization of total social net revenues is merely a didactic crutch. Present total social net revenues can be expressed, not graphically in terms of Dupuit's famous area, but only as an integral of a multivariable function. As in private economics, maximization can formally be accomplished by employing the joint-production approach or the calculus of variations.

It follows that care has to be exercised in interpreting the criteria for maximizing present social net revenues as equality of price and marginal social costs. In time economics, one must compare the present value of future marginal revenues occurring usually in more than one interval with the present value of future marginal costs likewise occurring in a number of intervals. Even if marginal revenues in different intervals are assumed to be independent, marginal costs are always interrelated. The main reasons are the effects of cumulative use and of sunk costs.[7]

[5] Hotelling has refined Dupuit's treatment through use of better mathematical tools; in a controversy with Frisch he has clarified the role of the marginal-cost concept: cf. footnote 1.

Ragnar Frisch, "The Dupuit Taxation Theorem," *Econometrica*, Vol. 7, No. 2, April 1939, pp. 145–150.

Harold Hotelling, "The Relation of Prices to Marginal Costs in an Optimum System," *Econometrica*, Vol. 7, No. 2, April 1939, pp. 151–155.

Ragnar Frisch, "A Further Note on the Dupuit Taxation Theorem," *Econometrica*, Vol. 7, No. 2, April 1939, pp. 156–157.

Harold Hotelling, "A Final Note," *Econometrica*, Vol. 7, No. 2, April 1939, pp. 158–160.

[6] J. R. Hicks, *Value and Capital.* (2d. ed.; Oxford: The Clarendon Press, 1946), pp. 38–41.

J. R. Hicks, "The Rehabilitation of Consumers' Surplus," *The Review of Economic Studies*, Vol. 8, No. 2, February 1941, pp. 108–116.

J. R. Hicks, "The Four Consumer's Surpluses," *The Review of Economic Studies*, Vol. 11, No. 1, Winter 1943, pp. 31–41.

Alexander Henderson, "Consumer's Surplus and the Compensating Variation," *The Review of Economic Studies*, Vol. 8, No. 2, February 1941, pp. 117–121.

[7] External economies and diseconomies are implicitly taken into account in social cost functions. If external diseconomies are caused by fixed factors in the resource

The difficulties of equating marginal social revenues and marginal social costs in time economics are relevant also for price fixing if marginal costs are to be used as a criterion.[8] The two problems, however, are not identical. In price fixing, price can be equated to current marginal social costs, and plant maintenance and expansion can be made subject to a separate administrative decision and financial appropriation. In planning an optimum time distribution of use rates, on the other hand, the present value of future marginal social revenues and costs in usually more than one—and possibly many—interrelated intervals must be considered.[9]

Social Revenues and Costs in Relation to Market Form

The preceding interpretation of marginal social revenues and costs points to a possible difference between private and social optima in the state of conservation. One necessary condition—besides several others to be discussed below—for an identity of private and social optima in the state of conservation is pure competition: rates of use and prices of products corresponding to an equality of marginal *social* costs and revenues is—theoretically—realized by private enterprises under pure competition. If monopolistic conditions, in the defined sense (chapter 5), prevail, there will generally be differences between the social optimum in the state of conservation and that state established by private enterprises.

How monopolistic conditions in resource use may affect the state of conservation was considered in chapter 14. Differences between the social and private optima in the state of conservation caused by the existence of monopolistic conditions are not confined to cases in which such conditions exist in the use of the par-

industry, or by increasing prices of services withdrawn from alternative employments, or by both, they are transfer items in the explained sense and should not be included in marginal social costs.

[8] Some of these difficulties are discussed by:

Emery Troxel, "Incremental Cost Determination of Utility Prices," *Journal of Land and Public Utility Economics*, Vol. 18, No. 4, November 1942, pp. 458–467.

Emery Troxel, "Limitations of the Incremental Cost Patterns of Pricing," *Journal of Land and Public Utility Economics*, Vol. 19, No. 1, February 1943, pp. 28–39.

[9] See Appendix, section 2.

ticular resource which is being studied. Monopolistic conditions elsewhere in the economy may affect the state of conservation decisively—for example, monopolistic conditions in the market for capital funds. As has been explained, the state of conservation is very sensitive to interest rates and capital rationing.

In the modern capitalistic economy, monopolistic conditions are widespread. It cannot be the objective of conservation policy to eliminate these conditions. Other public policies have broader antimonopoly objectives. Such policies cannot be discussed here. It was noted earlier that these policies have frequently not succeeded in restoring pure competition, but have merely discouraged pure monopoly in favor of oligopoly. Such results cannot be appraised with the tools of economics alone. Nevertheless, the all-pervasive effects of monopolistic conditions in the modern capitalistic economy are not without relevance for conservation policy. These effects restrict the validity of any argument for or against a particular conservation policy if the criteria are existing market values. Even more important restrictions on such criteria will be taken up next.

Allocation and Incidence of Social Revenues and Costs

It may be regarded as axiomatic for social planning (accounting)—at least under positivistic postulates (see p. 245)—that *all* revenues and costs of a production plan must be considered regardless of who receives or pays them.[10] It is well to differentiate clearly between such an imputation of revenues and costs to functionally related use as a problem of proper social accounting, and the actual receipt and expenditure by persons as a problem of income distribution. The former problem will be called here, for short, "allocation" and the latter "incidence." Allocation to individual use rates under joint production and the techniques of allocation have been discussed elsewhere.[11] At this point, attention is focused on allocation to individual production

[10] The Federal Flood Control Act of June 22, 1936 [49 *U. S. Stat. at L.* (1936), pp. 1570–1597] employs the phrase "to whomsoever they may accrue."

[11] S. V. Ciriacy-Wantrup, "Economics of Joint Costs in Agriculture," *Journal of Farm Economics*, Vol. 23, No. 4, November 1941, pp. 771–818.

plans. In this connection, differentiation between several portions of social revenues and costs is useful.

Social revenues and costs include all revenues and costs functionally related to a production plan *regardless of incidence;* most of our present discussion (Part IV) refers to them. Social revenues and costs are partly private and partly public.

The incidence of a portion of private revenues and costs is on the planning agent responsible for a production plan. This portion may be called "entrepreneurial" revenues and costs. The private economics of conservation is largely concerned with this portion (Parts II and III).

The incidence of a second portion of private revenues and costs is on individuals who are connected with a production plan in nonentrepreneurial capacity—for example, as landlords, creditors, or laborers.[12] The importance of this portion for conservation was mentioned when the problem of compensation for landlord and tenant was discussed.

The incidence of a third portion of private revenues and costs is on individuals under conditions where interenterprise connections exist. Such connections, for example, are neighborhood relations mentioned in our discussion of property rights. The second and third portions taken together may be called "private nonentrepreneurial" revenues and costs.

The incidence of public revenues and costs may be on individual users of specific public properties—for example, roads, schools, reservoirs, and parks; or it may be diffused among the people of a region, a state, or a country—for example, if effects on the levels of public education, public health, economic stability, and national defense are considered. The latter may be called "collective" revenues and costs.

---

[12] The interest of landlords and creditors in conservation is generally recognized. However, labor's stake in conservation is frequently not smaller. Labor is not perfectly mobile. Conservation may affect security of employment (stranded workers in forest areas not under sustained-yield management are an example) or conditions of housing (where housing is temporary because of resource depletion as in some mining areas) or personal health and comfort (through dust storms, water pollution, and decay of scenic resources).

Allocation and Incidence in Conservation Policy

Private planning agents usually take only entrepreneurial revenues and costs into account. Frequently not even these are fully considered. For example, in soil depletion the individual farmer is frequently unaware of all entrepreneurial costs. If there are discrepancies between the revenues and costs considered by individual planning agents and social revenues and costs, there will be differences between the private and the social optima in the state of conservation. In a free-enterprise economy, at least, such differences are one of the most important challenges for conservation policy. The objective of conservation policy is to reduce such differences by inducing private conservation decisions to approach the social optimum more closely.

The above differentiation between different portions of social revenues and costs is highly important for selecting the appropriate tools of conservation policy. For example, in order to induce proper (in the sense of allocation "as if" for social accounting) consideration of entrepreneurial revenues and costs, education is frequently sufficient; proper consideration of private nonentrepreneurial revenues and costs can frequently be induced by more vigorous application of existing property laws, by subsidies, or by several other policy tools discussed in chapter 19; for proper consideration of public revenues and costs—besides the same tools—outright public management may sometimes be necessary.

Social versus Entrepreneurial Revenues and Costs

Sometimes it is held that the public interest is not involved if entrepreneurial revenues and costs form a substantial part of social revenues and costs, as in many soil-conservation practices on private land. According to this view, such soil-conservation practices should not be subsidized by the public. One may ask, when is the public interest involved? If one, ten, or twenty per cent of social revenues and costs are nonentrepreneurial? If one, ten, or twenty people are involved in addition to the planning agent? Highly arbitrary procedures are sometimes adopted to "solve" this difficulty. The W.P.A., for example, adopted a general administrative rule to the effect that at least twenty people should share in the benefits of a public project before it could be approved.

The solution of this difficulty has already been indicated: The public interest is equally involved in all portions of social revenues and costs. However, it was also indicated that the tools for safeguarding the public interest may vary with respect to different portions and, within portions, with respect to causes of improper allocation. The mere fact that social revenues and costs are also largely entrepreneurial does not rule out a public interest. On the other hand, subsidies may not be at all socially desirable even if a substantial part of social revenues and costs are nonentrepreneurial.

If public subsidies for conservation are contemplated in cases where most revenues and costs are entrepreneurial, the following two criteria may be applied: First, we must ask, are subsidies the most economical tool for reducing differences between private and social optima in the state of conservation (considering, of course, costs of administration)? Or are alternative tools—for example, in the fields of education, taxation, and regulation—more economical? Second, assuming that subsidies are the most economical tool, are the resulting social net revenues greater than if the same amount of public money was spent in alternative fields? The economist is not absolved from considering these criteria for conservation policy by the fact that rules concerning them are sometimes established by precedent.

Extramarket Values and Conservation Policy

The next difficulty encountered in reinterpreting revenues and costs in social economics of conservation is created by the existence of extramarket values. This problem is analytically and practically quite different from that of allocation. There are, however, important relations between the two problems: it is more difficult for private planning agents to appropriate revenues or shift costs improperly (from the standpoint of social accounting) if market values are readily available, than if individuals affected and the courts encounter difficulties in valuation. Furthermore, in this and other occidental countries, social institutions determining the incidence of revenues and costs have developed with and around an individualistic market economy. Hence, they are not well adapted to a consideration of extramarket values—especially if the incidence of the latter is collective.

The problem of extramarket values was encountered earlier in the private economics of conservation. There, however, inclusion of extramarket values in revenues and costs did not create difficulties for defining in theory and approximating in actuality the optimum state of conservation. Private planning agents are able to compare subjectively utilities and disutilities connected with extramarket as well as market goods. The mechanism and the quantities of such subjective valuation are neither accessible nor relevant for economic analysis. For conservation policy, on the other hand, it is necessary to obtain some objective yardstick for comparing market with extramarket goods. Procedures relevant for such an objective valuation of extramarket goods may be called "administrative valuation." [13]

Administrative Valuation of Extramarket Goods

Administrative valuation can often use market values as starting points and bench marks. Courts, public utility commissions, and the like frequently operate on this basis. In conservation, many extramarket revenues and costs can be valuated in this way. For example, reduction of the siltation of reservoirs, of the pollution of streams, and of the impairment of recreational and scenic resources can be valuated to a large extent by using market values in auxiliary calculations. The particular kind of auxiliary calculation that is employed varies greatly with differ-

[13] The term "collective valuation," which occurs in the same connection [H. D. Dickinson, *Economics of Socialism* (Oxford: Oxford University Press, 1939), 262 pp.], is not employed here because market valuation is in a sense also collective, and because it might be confused with the same term used above in conjunction with revenues and costs. The term "administrative price," in the sense of a price resulting from administrative valuation, has no relation to the term "administered" price, which has been used to describe the rigidity of prices under monopolistic conditions.

ent problems. Discussion of the many kinds that have been used would lead us too far afield here.[14]

Instead of using market values in auxiliary calculations, market criteria—equality of supply and demand—may be used for administrative valuation. Students of socialist economics have given considerable attention to this type of valuation, to the setting and changing of administrative prices through trial and error. This type of administrative valuation is suited for individual but not for collective extramarket goods. Individual extramarket goods are divisible from the standpoint of consumers: they can be consumed by each individual at his own choice and in smaller or larger portions. Examples are enjoyment of parks and sport fields, where an entrance fee might be charged which can be adjusted in such a way that the facilities are fully used but not overcrowded.

The same type of administrative valuation may be used for individual goods that are, at present, valuated in a market but under market forms which do not lead to equilibrium (such as oligopoly and bilateral monopoly) or under market forms which, although leading to equilibrium, may be regarded as undesirable for other reasons (such as pure monopoly or monopsony and Chamberlin's monopolistic competition).

This possibility is not without significance for a market price which is of special importance for conservation—namely, the interest rate. As emphasized in chapter 7, the market for capital funds for individual resource users is frequently characterized by monopolistic conditions. If, in such cases, administrative valuation is applied to interest rates, existing market rates may be used as bench marks; or else equilibrium criteria—for example, stable prices or full employment—may be chosen.

Valuation of Collective Extramarket Goods

There remains usually a residual item not covered by types of administrative valuation which "lean" on market values

[14] In the Central Valley project studies, for example, the attempt was made to valuate the recreational facilities and opportunities created by the project. See U. S. Department of the Interior, Central Valley Project Studies, Problems 10 to 13. *Payments by Beneficiaries* (Washington: Govt. Print. Off., 1947), pp. 77–100.

or market criteria in the sense just discussed. This value residual consists of those social revenues and costs the incidence of which is collective. Examples are levels of public education, public health, and national security. The costs of producing these collective goods are to a large extent distributed among consumers through general taxation.

The essential problem for the administrative valuation of collective extramarket goods is to obtain a demand function—that is, a marginal social revenue function. A supply function—that is, a marginal social cost function—can usually be obtained from market values; if not, the principles of obtaining supply functions are the same as for demand functions. We may ask then how can demand functions for collective extramarket goods be obtained?

As stated in chapter 6, marginal rates of substitution between extramarket and market goods can sometimes be obtained objectively through observation of the individual planning agent's behavior in actual or hypothetical situations of choosing. If the market good is money, a schedule of such rates of substitution can be interpreted as an individual demand schedule for the extramarket good.[15] Such individual schedules can be arrived at by questionnaire, and out of these schedules a demand function corresponding to a market demand function can be obtained—subject to several possible objections to be considered presently.

Individuals of a sample or of a social group as a whole may be asked how much money they are willing to pay for successive additional quantities of a collective extramarket good. The choices offered relate to quantities consumed by all members of a social group. If the group interrogated is a sample, and only one sample is used, the modal schedule of the sample is obtained, and each point on this schedule is then multiplied by the number of individuals in the whole group being investigated. If every individual of the whole group is interrogated, all individual values (not quantities) are added.[16] The results correspond to a market

[15] As noted before in a different connection (chap. 7, footnote 5), marginal rates of substitution at a given indifference level can be interpreted as ratios between marginal utilities. These ratios are equal to the ratios between prices (marginal unit values) of the goods considered.

[16] In individual (in contrast to collective) extramarket goods, the quantity scale of individual demand schedules relates to individual units of the good. In order to

demand schedule. For purposes of public policy, this schedule may be regarded as a marginal social revenue function. In combination with a corresponding cost function, the socially desirable supply of the collective extramarket good can be determined.

Objections to This Valuation Procedure

Several possible objections to this valuation procedure may be considered.

First, we may object that the suggested transition from individual to market demand schedules is not feasible because individual schedules may be nonadditive. It is true that caution is advisable if individual demand schedules for individual (in contrast to collective) goods are added: each individual demand schedule may be dependent on (contain as variable) the total quantity demanded by all consumers or by a particular group of consumers. In more familiar terms, some individuals in some purchases will like to "keep up with the Joneses" and other individuals (or the same individual in other purchases) will like to "be different from the Joneses." This difficulty of nonadditivity is not present in our case because questionnaires relate only to the total quantity of the extramarket collective good available to all members of a social group. However, we have to consider a somewhat different point: Individual demand functions may be dependent on the number and personal characteristics (race, religion, sex, age, and so on) of other consumers who demand the same good. For example, people may be influenced in their willingness to pay for recreational facilities by their assumptions of how many and what kind of other people are going to use the same facilities. Generally this influence can be regarded as minor; if necessary, it can be taken into account in administrative valuation by indicating in the questionnaire definite assumptions—for example, with respect to the use of successive additional quantities of recreational facilities—and by varying these assumptions in repeated investigations.

Second, one may object that quantities of collective extramar-

---

obtain the total demand schedule (marginal social revenue schedule), quantities are added. In the case discussed in the text, values are added.

ket goods are characterized by lumpiness, and that marginal values and quantities, therefore, may be difficult to obtain and may not have much operational meaning. The problem of lumpiness is in no way peculiar to collective extramarket goods. It was suggested earlier that in practical approximations total additional rather than marginal values and quantities are used. It cannot be denied, on the other hand, that defining quantities of collective extramarket goods is difficult—regardless of their lumpiness. If no suitable physical quantities can be found, costs may serve; for example, one may measure additional conservation in terms of Congressional appropriations for this purpose. In this case it must be assumed that decisions with respect to priorities in the conservation effort are left to the proper administrative authorities. Still, the meaning of such appropriations in terms of physical quantities must be explained to the voters and legislators in order to enable them to make meaningful choices.

Third, one may object that expectations of the incidence of costs in the form of taxes will bias the responses to interrogation. Through proper education and proper design of questionnaires or interviews it would seem possible to keep this potential bias small. Guarding against such a bias is a necessity common to all census questionnaires which bear on income.

Fourth, one may object that marginal social revenue functions obtained in the way suggested cannot be used, since other extramarket goods are not considered, and since marginal utility of money is not likely to remain constant. These objections apply also to the use of demand functions in analyzing the market. Strictly speaking, of course, *all* market and extramarket goods are interrelated because of complementarity or competitiveness in consumption or production and because of limited money income. Theoretically one can obtain, through interrogation, marginal social revenue functions for any number of variables; changes in marginal utility of money can also be measured and accounted for. In economic reality, public policy must usually be content to deal with one extramarket good at a time and to assume that changes in the marginal utility of money are of the second order of smalls. This assumption is frequently realistic because of compensating variations in the prices of other commodities or in money income.

Finally, the proposed method of valuating marginal social revenue functions of collective extramarket goods may be rejected as too academic. It is true, of course, that the supply of collective extramarket goods is usually not determined by monetary valuation, but by the political machinery of a democracy, the command of an autocrat, or in other ways. This does not mean, however, that in a democracy, objective valuation might not aid the government in making decisions. It may be noted that the procedure outlined can be used for quantitative determination as well as for ranking of values. Considerable progress has been made recently in designing and evaluating group interrogations by questionnaire and interview. Economists, so far, have made little use of this progress in the field of individual and social psychology. Welfare economics could be put on a more realistic foundation if a closer coöperation between economics and certain young branches of applied psychology could be established.

Weighting of Individual Preferences

The problem of extramarket values creates considerable difficulties for determining the optimum state of conservation in social economics. Still, these difficulties are not basic for conservation policy. They can be reduced by procedures already suggested and by approximations discussed below. There are other problems, however, which present a more basic difficulty for conservation policy. These problems impose irreducible restrictions on the validity of objectives and criteria of conservation policy which are defined in value terms. One of the problems we have in mind is that of social weighting.

In market valuation, social weighting is accomplished by the distribution of money income. In this case the social weights are attached to the individual, and he is allowed to distribute his weight among goods and services in influencing market prices. In administrative valuation weighting is partly by income (as far as market values are used as bench marks) and partly by the distribution of political rights, that is, by the influence of individuals upon the administrators. Practically, administrative weighting is possible only in a rather general way for whole groups of individuals and for whole groups of goods and services. An example is rationing in wartime, when soldiers, heavy workers, normal consumers, children, and other groups may administratively receive ranks (priorities) in the distribution of certain foods.

The normative aspects of social weighting cannot be approached with the tools of economics. The social significance of individuals and of their preferences depends on the whole culture of which economics is only a part. A particular system of social weighting may, however, be selected for economic study because it is characteristic of a certain society or because it is postulated on philosophic, ethical, or other *a priori* grounds.

Postulates Concerning Social Weighting

In this study, equality in the *opportunity* to acquire income and to exercise political rights may be selected because it is the professed ideal of contemporary democratic societies, and because the selection is suggested by the philosophic attitude known as positivism. The first reason implies that some practical significance may be claimed for the selection. The second reason not only contains the logical justification of democracy, but is also the basis of the scientific approach to social problems. If any metaphysical absolutism of values and of truth is rejected, as in positivism, there is no *a priori* basis for differentiating between the social significance of individuals.

There is no need to emphasize that practical application of a professed equalitarian ideology in contemporary democracies is, to say the least, highly incomplete. If equality of opportunity has existed in historical societies, it did so only for short periods. There is no evidence that this situation is greatly different in a socialist society, and no convincing logical reason why it should be—except that the differentiating effects of property and birth are less. Individuals entrusted with the power of government on the basis of socialist institutions cannot well be prevented from forming a ruling group and from becoming an economically privileged class. Continuity of a privileged class need not be based on birth. The Catholic clergy in the Middle Ages and the party hierarchy in modern totalitarian states are examples in which a class tries to preserve its continuity through selection and training rather than through blood relationship.

Distortions of the Optimum State of Conservation

The objective of conservation policy is not to obtain equality of opportunity. Moreover, as just stated, removal of inequality of opportunity offers little prospect of accomplishment. The objective of conservation policy is limited, therefore, to compensating for the "distortions" (under our postulates) in the optimum state of conservation caused by unequal opportunity.

Inequality of income based on inequality of opportunity affects market valuation of products, of productive services, and of interest rates. This may mean either conservation or depletion as compared with the state of conservation that would exist if incomes were more equal. Compensation through conservation policy must vary accordingly. Three steps are required for such compensation: First, the effects of variations in the income distribution upon prices and interest rates must be ascertained, other variables, including aggregate money incomes, being held constant. Second, the effects of such changes in prices and interest rates upon the state of conservation must be analyzed. Third, the state of conservation must be changed in the direction of whatever state would exist if the income distribution resulting from our postulates should prevail.

Compensating Conservation Policy

Some general indications for shifts of demand schedules under the influence of changes in the income distribution can be obtained by studying consumers' expenditures in different income groups, particularly the sign and the form of regression between income and consumption expenditure.[17] Probably, for

[17] U. S. National Resources Board, *Agricultural Land Requirements and Resources,* Part 3 of the Supplementary Report of the Land Planning Committee (Washington: Govt. Print. Off., 1935), p. 9.

U. S. National Resources Committee, *Consumer Expenditure in the United States*

example, a more equal income distribution than the existing one would lead to upward shifts of demand schedules for dairy products, eggs, meat, fruits, and leafy vegetables, with little change in demand schedules for grains, potatoes, sweet potatoes, and cotton. A corresponding shift in land utilization away from grains and cotton toward livestock, fruits, and vegetables would, as a whole, lead to conservation of soil resources for technological reasons. On the other hand, such a study suggests that a more equal distribution of incomes might lead to decreases of aggregate savings. This would tend, other things being equal, to an increase in the market rate of interest and to depletion. According to our discussion in chapter 7, however, a more equal income distribution may lead to large and numerous decreases (with fewer and smaller increases) of individual time-preference rates. Under impure and imperfect markets, individual utilization plans may be determined by individual time-preference rates rather than by market rates of interest.

An attempt has been made to measure statistically the influence of changes in the income distribution upon consumption by holding constant aggregate incomes and other variables such as time and cost of living.[18] That investigation obtains negative coefficients for Pareto's a [19] in a multiple correlation with expenditure

(Washington: Govt. Print. Off., 1939), 196 pp. In Part 2, Section 6 of this work, references to previous pertinent studies are given.

U. S. Bureau of Home Economics, *Family Food Consumption and Dietary Levels, Five Regions* (Washington: Govt. Print. Off., 1941), 393 pp. (U. S. Department of Agriculture Miscellaneous Publication 405, Consumer Purchases Study: Farm Series).

U. S. Bureau of Home Economics, *Family Food Consumption and Dietary Levels, Five Regions* (Washington: Govt. Print. Off., 1941), 268 pp. (U. S. Department of Agriculture Miscellaneous Publication 452, Consumer Purchases Study: Urban and Village Series).

U. S. Bureau of Labor Statistics, *Family Expenditures in Selected Cities, 1935–1936* (Washington: Govt. Print. Off., 1941), 8 vols. (U. S. Department of Labor Bulletin No. 648, Study of Consumer Purchases: Urban Technical Series, Vols. 1–8).

U. S. Bureau of Home Economics, *Family Spending and Saving as Related to Age of Wife and Age and Number of Children* (Washington: Govt. Print. Off., 1942), 126 pp. (U. S. Department of Agriculture Miscellaneous Publication 489).

[18] J. J. Polak, "Fluctuations in United States Consumption, 1919–1932," *The Review of Economic Statistics*, Vol. 21, No. 1, February 1939, p. 4, Table 2.

[19] Polak defines a in such a way that the greater a is, the more equal is the distribution of income. For Pareto's view see Vilfredo Pareto, *Cours d'Économique Politique* (Lausanne: F. Rouge, Libraire-Editeur, 1897), Vol. 2, Book 3, pp. 299–345.

for consumption; the results were rejected as unreasonable, and only positive coefficients were regarded as acceptable. Changes in time-preference rates just mentioned may to some extent explain the results.

Empirical evidence is even less, and theoretical obstacles are greater when one attempts to ascertain the distortion of administrative valuation through existing inequalities of opportunity to exercise political rights. Large groups of individuals (for example, people below voting age, women in many countries, Negroes in the United States) have less weight per capita in influencing administrators than other groups. Lacking empirical studies of preferences of these groups, one can make no definite statement about the direction and the degree of this influence upon the state of conservation.

We must conclude that no definite suggestions regarding a compensatory conservation policy under equalitarian postulates can as yet be made. More data must be collected about the effects of changes in income distribution upon prices and interest rates and about the preferences of politically underprivileged groups. Conservation policy will then be able to view values as parameters under the influence of income distribution and other social weights. Until this aim is accomplished, measures of conservation policy must be satisfied with value criteria that are based on the existing social weights. However, the implied restrictions on the validity of such criteria should not be forgotten in conservation policy.

The Problem of Consumer Sovereignty

The preceding discussion is based on the assumption that the consumer is the independent, ultimate arbiter of a value system. Even administrative valuation in the above sense is directed by consumer's preferences. This seemingly so simple assumption of consumer sovereignty is usually mentioned quite incidentally, if at all, in economic discussion. However, it raises the most complex problem for public policy.

In many situations, the consumer is no more an independent sovereign than the weak ruler of a modern constitutional monarchy with a strong parliament and cabinet. In the actual world, the consumer's alternatives are not merely limited by monopolistic arrangements, as mentioned above, nor are they merely limited by laws; consumer's preferences are influenced to their very roots by the many means of education and propaganda in the hands of producers, traders, and, especially, the modern state. How can consumer's preferences be a guide for public policy if these preferences can be influenced to a considerable degree by public policy itself?

It must be recognized frankly that consumer's preferences and public policy are interrelated variables either of which may be the dependent one for the purpose of analysis. For our present purpose it is admissible to hold one of them constant—that is, to use a value system based on given preferences as a criterion of public policy. This procedure, however, imposes an additional severe restriction on the validity over time and place of such a criterion.

Practical Approximations of the Social Optimum

Enough has been said to show that theoretical refinement of the formal definition of the social optimum in the state of conservation—as that state in which social net revenues are maximized over time—is of only limited practical interest. It seems more appropriate to focus on some practical approximations to the social optimum in the state of conservation.

Much as in private planning, the practical goal in conservation policy is not the optimum time distribution of use rates, but a step-by-step improvement of the existing one through trial and error. The improvement is made by comparing the present value of total additional social costs of whole measures of conservation policy (changing use rates in a number of intervals) with the present value of total additional social revenues of such measures. In other words, the practical criterion is an increase of present total social net revenues.

The Social Optimum and Uncertainty

A still more radical reformulation of the goal of conservation policy is required by uncertainty of expectations. This problem likewise was encountered in private economics; it is even more important in social economics.

For some important resources (flow resources with a critical zone) allowance for uncertainty in social economics leads to an objective for the state of conservation which can be fairly well defined, and which in many cases is the most practical approximation to the social optimum. This approximation was previously mentioned as a "safe minimum standard of conservation"; it deserves a more detailed discussion.

CHAPTER 18

A SAFE MINIMUM STANDARD AS AN OBJECTIVE OF CONSERVATION POLICY

The Social Risk of Irreversibility

When the objectives of conservation decisions in private economics were discussed, an important influence of uncertainty was noted: in striving for the economic optimum, planning agents are frequently confronted with choices between possible losses of various magnitudes and probabilities. Generally, the possibility of larger but less probable losses can be avoided only by accepting the possibility of smaller but more probable ones. The practical means of accomplishing this were discussed under flexibility, hedging, pooling, and spreading. As a special case (but an important one), planning agents are interested in avoiding the possibility of the losses which in chapter 6 we called "immoderately" large, those contingencies which would entail bankruptcy—threaten the continuity of the utilization plan.

A similar situation prevails in social economics. Governments also are confronted with choices between possible losses of various magnitudes and probabilities. The possibility of immoderate losses—contingencies which threaten the continuity of a social group—are even more important for government decisions than for private ones. One such contingency in the field of conservation is the economic irreversibility of depletion in that important class of flow resources which is characterized by a critical zone. This class contains soil, water, plants, and animals. Why does irreversibility in the depletion of these resources carry the possibility of an immoderate social loss?

From a certain degree onward—for example, if more and more acres of land or species of plants and animals are affected—irreversibility in the depletion of critical-zone resources limits opportunities of adaptation and narrows the potential development of a society. Both the biological and the social sciences have come to the conclusion that such a limiting and narrowing force directs development toward specialization rather than diversification. Such a direction has been held responsible for retarded and abortive growth—in the sense of growth toward a dead end—stagnation, and death of species and civilizations. Examples for such results in the field of genetics have been pointed out by Dobzhansky [1] and his co-workers. In the field of social studies this theme has been emphasized by Toynbee, Kroeber, and others.[2]

It was emphasized in chapter 3 that economic irreversibility of depletion is uncertain, since it depends, at a given time and place, on "presently foreseeable conditions" in technology, wants, and social institutions. These conditions are in constant change. There is an additional uncertainty in the present context—namely, whether such irreversibility will actually lead to immoderate social losses. In other words, we are not considering reliable relations of cause and effect but a contingency to be guarded against: There is no certainty that depletion will actually prove economically irreversible in the future—although this out-

---

[1] Theodosius G. Dobzhansky, *Genetics and the Origin of Species* (2d. ed., rev.; New York: Columbia University Press, 1941), 446 pp.

[2] Arnold J. Toynbee, *A Study of History* (London: Oxford University Press, 1935–1939), 6 vols.

Alfred J. Kroeber, *Configuration of Culture Growth* (Berkeley and Los Angeles: University of California Press, 1944), 882 pp.

come may seem probable under presently foreseeable conditions. Furthermore, there is no certainty that economic irreversibility —if and when it actually does occur—will be a necessary or sufficient condition or "cause" for the stagnation or disintegration of a social group; this outcome may well have a rather small probability (although, according to some serious students, it has actually occurred in the past).

A decision to avoid the social risk of irreversibility is not dependent on whether or not the losses which threaten are immoderate. As we know, avoiding the possibility of immoderate losses is merely a special case of making choices between the possibility of larger but less probable losses and that of smaller but more probable ones. If the more probable losses are small in relation to the less probable ones which may be avoided by accepting the former, the economic choice between the two alternatives would not be difficult. What, then, are these smaller but more probable losses?

The Safe Minimum Standard of Conservation

They are connected with maintaining what we have called a "safe minimum standard of conservation." In the resource class under consideration, a safe minimum standard of conservation is achieved by avoiding the critical zone—that is, those physical conditions, brought about by human action, which would make it uneconomical to halt and reverse depletion. A safe minimum standard of conservation involves losses if its maintenance necessitates costs (either use foregone or positive efforts) and if the contingency guarded against should not actually occur—that is, if depletion should not prove economically irreversible. These losses are similar to the costs of flexibility in private economics. The similarity is more than formal: as implied above, a safe minimum standard of conservation is essentially an increase of flexibility in the continuing development of a society.

It was suggested in chapter 3 that critical-zone resources are the most important ones from the standpoint of the economic and social problems created by depletion. Conservation policy is mainly concerned with this class. Thus, a safe minimum standard is of great significance for conservation policy, although it applies only to one particular class of resources.

The critical zone may be defined in terms of a certain flow rate and of some corresponding use rate. In this sense, the safe minimum standard is a "state of conservation" according to the definition of chapter 4. However, actual maintenance of any particular use rate is not the primary objective of a safe minimum standard but merely a by-product. The primary objective is to maintain the economic possibility of halting and reversing a decrease of flow and use.

Let us ask, then, how large—absolutely and relatively—are the costs of maintaining the safe minimum standard of conservation?

Costs of Maintaining a Safe Minimum Standard

The costs of maintaining a safe minimum standard are absolutely small if proper action is taken in time and if the proper tools of conservation policy are employed. These two conditions will be discussed in detail in the following chapters. At this point, we will proceed to discuss the costs of maintaining a safe minimum standard of conservation under the assumption that these two conditions are realized.

In *some* practical situations, maintenance of a safe minimum standard necessitates that use is foregone. Use rates in the neighborhood of the critical zone are small and, in the alternative case —that is, if a safe minimum standard were not maintained—could continue over only a small number of intervals. It is well to remember that the safe minimum standard of conservation is more modest than a theoretical social optimum. Frequently, such a standard corresponds to a state of conservation which is considerably lower than even the private optimum. Many private enterprises, frequently the majority, may be operating above the safe minimum standard, on the basis either of economic calculation or of habit patterns.

In *many* practical situations, maintenance of a safe minimum standard does not involve any use foregone; rather, it involves a change in the technical ways (not the quantities) of utilization. Such changes may be brought about by the tools of conservation policy. These changes may or may not necessitate costs in the sense of positive efforts (inputs) by individual planning agents or by the public or by both. Sometimes a change of social institutions without any inputs is sufficient. Sometimes the costs are only public—for example, if education or a temporary subsidy is the most economical tool of conservation policy. If private costs are increased—for example, as a consequence of regulation by governments or public districts—only a few enterprises may be affected, because, as emphasized above, the minimum standard of conservation is a rather modest objective even in terms of the private optimum.

With proper timing and choice of tools (the two conditions mentioned above), costs of maintaining the safe minimum standard are not only small in absolute amount, but very small relative to the loss which is being guarded against, a decrease of flexibility in the continuing development of a society. Costs of maintaining the safe minimum standard of conservation are also very small as compared with generally accepted expenditures of a social group for safeguarding its continuity in other fields. Such fields are, for example, public health and safety and national defense. In these fields, likewise, a safe minimum standard is frequently adopted as an objective of public policy. The reason is the same as in the field of conservation: It is impractical to determine the social optimum in the state of public health and safety or national defense because of uncertainty and because of the difficulties of valuating social revenues and costs. On the other hand, it is practical to set up standards which would avoid serious losses—threats to social continuity—in cases of epidemics, internal disorder, and foreign military involvements.

Whether costs of maintaining a safe minimum standard of conservation are small absolutely and very small relatively is a question of fact. It is necessary, therefore, to spell out in more detail the safe minimum standard for the various flow resources which are characterized by a critical zone in their depletion. We will attempt to do this in two stages: First, we will try to get a clearer idea of the physical conditions—brought about by human action—which are involved in the economic irreversibility of depletion for important flow resources. Secondly, we will try to define—for the practical purpose of conservation policy—a safe minimum standard applicable to these physical conditions.

The Critical Zone

The physical conditions which characterize the critical zone are comparatively simple for individual species of animal and plant life: with the destruction of the breeding stock or, frequently, its natural habitat,[3] flow becomes economically (and technologically) irreversible.

Depletion of ground water becomes economically irreversible through compaction and certain forms of pollution. The former may occur as a consequence of sustained overdraft; the latter may occur through changes in the salt balance and direct saltwater intrusion, also made possible by overdraft, or if the protection by impervious strata separating a usable aquifer from a chemically unusable one is destroyed through improper drilling and other activities.

Soil depletion may become economically irreversible if a protective plant cover is destroyed by cultivation, improper degree or timing of grazing, and repeated burning in areas where topography, climate, or soil render precarious any existing balance between erosive forces (water, wind) and the stabilizing force of plant cover. Erosion, even if it becomes irreversible, is somewhere offset by deposition. Sometimes, deposition by floods is an economic gain—for example, if fertile silts are deposited or if movement and deposition of coarse materials are necessary for maintaining infiltration areas to replenish ground water. More often, however, such deposition—in rivers, reservoirs, harbors, on cultivated flood plains—is an additional social loss rather than a gain. Furthermore, acceleration of the processes of erosion and deposition may make another important resource, surface water flow, unusable for many purposes (drinking, bathing, fishing, and some industrial uses).

---

[3] Frequently a breeding stock cannot be maintained in botanical or zoölogical gardens—sometimes the cheapest way—if highly complex biological relations are involved. In such cases, appropriate reserves must be established and protected.

In the depletion of cultivated agricultural soil—in contrast to the soil under forests and permanent grasses—erosion may become economically irreversible at a rather early stage if it proceeds in the form of gullies. Gullies quickly make farming operations uneconomical.

In the depletion of forests and range, another physical condition—besides soil erosion—must be considered in connection with economic irreversibility. Forests and grasslands are plant (species) associations of often great complexity and sensitivity with respect to influences which upset the ecological balance. Repeated burning and improper degree or timing of grazing may upset this balance to such an extent that the valuable species are replaced entirely by other less valuable ones. Such a degeneration of plant associations has occurred, for example, on some cutover forest areas in the northern Great Lakes states, in the arid grasslands in the intermountain region, and in some mountain meadows of California. In such cases there may be a good plant cover and no serious soil erosion. Still, serious depletion has occurred which may be economically irreversible.

Depletion of recreational resources may become economically irreversible if the specific recreational uses depend on keeping an area "unspoiled." Spoliation may be caused by improper construction of access roads, permanent structures (hotels, dams), cutting of selected groves of trees (especially species which require a long time to reach impressive recreational value, such as redwoods and sequoias), man-made erosion (overgrazing of mountain meadows), and other effects of use.

Practical Definition of a Safe Minimum Standard

The great variety and complexity of physical conditions which characterize the critical zone in the depletion of the various flow resources make it generally impractical to define a safe minimum standard for each resource simply in terms of a single flow rate which is to be maintained. It is more practical to define a safe minimum standard in terms of conservation practices designed to avoid the critical zone. Such a definition may be in terms of conditions to be maintained—that is, in terms of the results of a number of unspecified conservation practices (definition in terms of results), or in terms of the performance of specific conservation practices (definition in terms of performance). A few illustrations may be helpful.

Definition in terms of results prevails in the following cases: In soil conservation, a safe minimum standard may be defined as the avoidance of gullies or as a maximum rate of erosion (in cubic feet per acre or acre-feet per square mile per year). In forest conservation, a safe minimum standard may be defined as a maximum rate of burn (in per cent of total area per year) or as maintenance of a given plant association; through the study of plant indicators modern ecology has made it practical to define a plant association and to check its maintenance periodically. In the conservation of grazing lands, a safe minimum standard may be defined in terms of a certain minimum amount of plant material (in tons per acre per year) left aboveground after the grazing season or, as in forest conservation, in terms of maintenance of a certain plant association. In the conservation of a plant or animal species, a safe minimum standard may be defined in terms of maintaining a certain breeding stock or in terms of protecting a certain area of natural habitat. In the conservation of water resources, a maximum degree of pollution (in terms of total or specific solids, bacterial count, oxygen conditions, and so on) may be used for defining a safe minimum standard.

If we now turn to definitions in terms of performance of specific conservation practices, we may first of all remember that mere limitation of use may be an important conservation practice. Frequently, therefore, a safe minimum standard may be defined simply in terms of a maximum use rate—"safe yield" (in acre-feet per year for a certain aquifer) in ground-water conservation, maximum rate of stocking (in animal-unit-months per acre per year) in the conservation of grazing lands, maximum annual harvest in hunting and fishing, maximum admission of people to avoid overcrowding of recreational facilities.

However, limitation of use is only one among many conservation practices. Particularly in agriculture and forestry, a maximum use rate is not very practical because use consists of a highly complex composite of products. These maximum use rates would have to be defined in terms of some common denominator (calories, digestible nutrients, dry matter, tons weighted by value, and the like) of various products. Other conservation practices—contour cultivation, mulching, strip cropping, terracing, leaving of a minimum number of seed trees, removal of slash and many others—are usually more practical. Definition in terms of conservation practices other than a maximum use rate may be employed even in fields where the latter is practical—rotation of pastures in grazing, proper capping and perforation of wells in ground-water pumping, specific processes in the treatment of polluted surface waters, prohibition of certain methods in taking game and fish.

All these practices are merely examples. Others may be better suited to particular practical situations.

Advantages and Disadvantages

The main advantages of defining a safe minimum standard in terms of conservation practices are great adaptability to local conditions, easy understanding by resource users, and fairly economical administration by governmental agencies charged with the execution of conservation policy.

The main disadvantage is that the suitability of conservation practices for avoiding the critical zone in resource depletion must be established first. This takes time and a considerable amount of observation and analysis. For establishing such suitability, it is not sufficient to show that a given conservation practice or combination of practices is technologically effective in avoiding the critical zone. It must also be shown that no other practice or combination of practices accomplishes the same result more economically. Adoption of a safe minimum standard as an objective of conservation policy does not mean that economic calculation can be neglected: a safe minimum standard of conservation should be realized with minimum total social costs.

The administrative advantages of defining a safe minimum standard in terms of conservation practices will probably lead to the employment of this method not only in agriculture and forestry, where it appears to be the only practical one, but also in cases—such as wildlife and range conservation—where a definition would be feasible which is more directly related to the maintenance of minimum flow rates. This trend in itself is not objectionable provided the relations between conservation practices and maintenance of a minimum flow rate are carefully analyzed, not only in their technological but also their economic aspects.

A Safe Minimum Standard and Private Enterprise

For many flow resources in many regions of the world, at various historical periods, the state of conservation established by individual resource users has fulfilled the requirements of a safe minimum standard without any influence from conservation policy. Such a state of conservation may be established on the basis of either economic calculations or habit patterns or both.

In northern and central Europe, at the present time, most agricultural soils, forests, and grasslands are managed by private individuals on the basis of economic calculations with an optimum state of conservation far higher than the safe minimum standard. Possibly the same would be true for the management of water and recreational and wildlife resources if their conservation were left to private economic calculations. In these resources, however, the problem of allocation of social revenues and costs seemed so important that the people have not taken the risk of leaving the maintenance of a safe minimum standard entirely at the mercy of private economic calculations.

In some parts of the world, habit patterns rather than private economic calculations or public policy have established a state of conservation above the safe minimum standard. Examples were mentioned in chapter 8. In such cases a state of conservation which would be established on the basis of private economic calculation might be lower than the state of conservation which exists on the basis of habit patterns—possibly even lower than the safe minimum standard. Where that is true, a conservation policy which relies on educating resource users to employ a more calculative attitude and more refined systems of accounting would defeat its own purpose.

In the United States, by far the largest portion of agricultural, forest, and grazing lands are managed by private individuals with a state of conservation at or above the safe minimum standard. Mostly this is due to economic calculation; but habit patterns, traditional standards of good husbandry, are by no means irrelevant even in a country where economic calculativeness is generally held in great social esteem. Adoption of a safe minimum standard as an objective of conservation policy therefore does not interfere with the practices of the great majority of private resource users. For the rest, a safe minimum standard would involve a minimum of restrictions both in geographic extent and degree. This limitation with respect to interference with private initiative and control may be regarded as a distinct advantage of adopting a safe minimum standard as compared with the adoption of any higher state of conservation, whether based on some notion of a social optimum, preëxisting conditions, or the *status quo*. Whether or not this particular advantage is accepted as such, constitutes, of course, a value judgment on other than economic grounds.

The Safe Minimum Standard as an Economic Base Level

The statement just made—that by far the largest portion of agricultural, forest, and grazing lands in the United States are managed by private individuals with a state of conservation at or above the safe minimum standard—does not necessarily conflict with the surveys of several government agencies and many individual writers. Such surveys suggest that most soils, forests, and grasslands of the United States are still being seriously depleted and, in many individual cases, could be utilized with greater private profit if conservation practices were applied.

Adoption of a safe minimum standard does not mean that public conservation policy should not be concerned with these conditions of depletion and insufficient information. A safe minimum standard of conservation should be generally adopted as a social objective and should actually be realized under all conditions as a kind of economic base level in conservation policy. As this chapter attempts to show, the economic rationale for adopting a safe minimum standard defined in physical terms is allowance for uncertainty; actual realization of a safe minimum standard is subject to the economic requirement of minimum total social costs.

After a safe minimum standard is assured, conservation policy should explore whether and how the social optimum in the state of conservation could be more closely approximated in the step-by-step fashion discussed in the preceding chapter. Thus, from a safe minimum standard of conservation as the base level, the economic advisability of further conservation practices may be calculated by comparing additional total social revenues and costs.

Safe Minimum Standard and Resource Relations

The safe minimum standard of conservation as an objective of public policy applies to the resources of one particular class (class II, 2, *b*). It may be asked what is the effect upon the economic rationale of such a standard if relations between these resources and others in different classes are involved? Such relations were classified as complementarity, competitiveness, or neutrality in supply and in demand.

The economic rationale of a safe minimum standard is based on the proposition that the costs of maintaining it are small in relation to the possible losses which irreversibility of depletion might entail. Complementarity in supply or in demand or in both means that the costs of maintaining the safe minimum standard are decreased or the losses of irreversibility are increased or both, as compared with that situation which would prevail if relations between resources could be neglected in the analysis (neutrality). Thus, complementarity between resources strengthens the argument for maintaining a safe minimum standard; for this reason we need not discuss the problem of complementarity further. The opposite—that is, a weakening of the argument for a safe minimum standard—would seem the consequence of competitiveness; this problem, therefore, must be explored further. First we may take up competitiveness in demand and then competitiveness in supply.

Competitiveness in Demand

If critical-zone resources were competitive in demand with (could be replaced by) other flow resources (those without a critical zone) or with abundant stock resources, then the possible losses which irreversibility might entail could not be very great. That such competitiveness in demand prevails is implied in some speculations concerning the world's future food supply.

It has been asserted that industrial photosynthesis will become feasible on a large scale (either based on lower organisms—unicellular algae—highly efficient in photosynthesis, or on synthetically produced chlorophyll), and that the resulting carbohydrates will be transformed into proteins and fats through certain yeasts already available and by other techniques under experimentation.

Let us assume that these transformations will become actual fact, that the products will contain not only calories and proteins but all the necessary nutrients, and that they will be no less digestible to man and no more expensive than the products now derived from plants and animals. Even under these heroic assumptions, the resources under consideration here—plants, animals, soil, clear stream flow—would still remain vital: Man does not live by bread alone. The present degree of synthetic living—small as it is compared with the degree just assumed—has already set in motion strong counteracting tendencies in human needs. These needs in some form or other drive man "back to nature." They can be expected to increase greatly if plants, animals, and soil should ever be replaced by the laboratory and the factory as sources of calories, proteins, and other components of human nutrition.

In that case, these critical-zone resources would still remain unique to satisfy the demand for recreation (in the true etymological meaning of the word), for aesthetic enjoyment, for exercising all those physical, mental, and emotional capacities of man not needed any more for obtaining his daily bread but still necessary for his happiness, and necessary for his culture, as we can think of it, to survive. Culture in all its forms has developed in close interaction with plants, animals, soil, and clear stream flow. In different words, the demand for these resources is not based solely on their use in nutrition. In other, even more important, uses, competitiveness in demand with resources in other classes does not exist.

Competitiveness in Supply

Let us now turn to competitiveness in supply. If relations between resources are competitive in supply, the question arises whether maintaining the safe minimum standard in one resource might not make it economically impossible to utilize the related one. Such an outcome would defeat the social objective of a safe minimum standard, namely to avoid a narrowing and specialization in the resource base of a society. Cases in which such an outcome appears plausible at first sight, were mentioned in chapter 3—for example, the competition of agricultural production with coal production in the strip coal mines in several eastern states and the competition of grazing with placer gold mining in many alluvial valleys of California.

Competitiveness—like complementarity and neutrality—depends on the numerical values of use rates. As has been emphasized, the safe minimum standard is a very modest one in this respect. The safe minimum standard does not require that use is maintained uninterruptedly, but that use may be resumed at some future date. An uninterrupted (small) use is merely a frequent by-product of maintaining the safe minimum standard (p. 254). Thus, a safe minimum standard in one resource is usually compatible with the utilization of a resource which is competitive in supply, although any higher state of conservation would not be compatible. Let us consider a practical illustration:

A safe minimum standard in soil conservation does not require that after strip or placer mining the land must be restored to such a condition that the previous use could be resumed at the same level immediately.[4] If such a requirement did exist, mining would become uneconomical privately as well as socially.

[4] It may be noted that the present value of net revenues from utilizing the stock resource (coal, gold) is invariably much greater than the market value (price) of an

The requirement of a safe minimum standard is more modest: the land must be restored to such an extent that under presently foreseeable economic conditions resumption of the previous use at some level at some future date is not impossible. In practice, this means the requirement that spoil piles are releveled and also revegetated (revegetation need not necessarily involve the previously existing species and need not necessarily be done by artificial means). The natural processes of soil formation will do the rest—provided that spoil piles do not contain toxic substances that make impossible all types of plant growth; the latter situation is comparatively rare.

In some states, such a minimum standard of soil conservation is imposed on strip-mining companies by law. One state—West Virginia—goes further: it requires that soil removed from coal beds (the overburden) be replaced. More will be said about the actual implementation of a safe minimum standard when soil-conservation legislation is discussed.

Conservation Policy If the Critical Zone Has Been Passed

The question may be raised what conservation policy should do in those cases in which the critical zone has already been passed. For example, in large sections of the loess areas of China and in some smaller sections of the Piedmont area in the southeastern United States, gullying has made it uneconomical to restore farmlands to their previous uses for cultivated crops.

On practically all of these lands the critical zone for other, "lower" uses—for example, grazing, forestry, production of game—has not yet been reached. Conservation policy should be concerned with establishing a safe minimum standard for these uses but not *necessarily*—as a base level in the terminology used above—with restoring such a standard for the previous use (cultivated crops). This course is implied in the definition of irreversibility. Such a definition, as we know, is in economic and not in physical terms. If the critical zone has been passed for a

---

acre of land on the basis of its use in agriculture, forestry, or grazing. However, this relationship of values, in itself, does not make a safe minimum standard of soil conservation irrelevant as a *social* objective.

certain use, the horse has been stolen, and no effort should be wasted in safeguarding it.

Safe Minimum Standard and Natural Changes of Flow

Finally, to avoid any misunderstanding, it may be well to repeat that the critical zone and, accordingly, the safe minimum standard as an objective of conservation policy, apply to changes of flow brought about by human action. It has already been emphasized that flow, in the absence of human action, need not be constant or infinite. Many natural [5] changes of flow of critical-zone resources exist and, in some cases, the rate of such changes is fairly well known. Such natural changes are connected, for example, with geologic erosion by water and wind, climatic changes, changes of ocean currents, volcanic activity, mutation of genes, and many others. It is perfectly possible that such natural changes decrease a flow irreversibly. There is no implication in our use of the concept "minimum standard of conservation" that avoidance of such an irreversibility is necessarily an objective of conservation policy.

On the other hand, the limitation of the concept just reëmphasized does not mean that natural changes of flow could not (technologically) or should not (economically) be affected by human counteraction or, more specifically, are not the concern of conservation policy. It is desirable to investigate carefully how far changes of flow are due to natural and how far to human causes. Such a differentiation may be relevant for selecting the proper tools for an attack both in the field of technology and policy. However, even if human action is not involved as a cause, this *in itself* does not mean that nothing could or should be done about it. Whether and what counteraction is to be taken is a problem of economics not essentially different from determining the optimum state of conservation in those cases in which a decrease of flow is caused by human action. The difference is that in the former case a safe minimum standard as a base level for conservation policy does not necessarily have an economic rationale—

[5] The term "natural" in this connection was defined in the beginning of this study as taking place without human action or influence (chap. 1, footnote 9).

because the costs of maintaining it may be excessive—whereas in the latter case such a rationale exists.

A few examples may serve to illustrate the foregoing statement.

In many arid areas of the world, wind has had and still has major effects upon the soil formation and movement and plant growth (for example, in the dust bowl of the High Plains, the southern San Joaquin Valley in California, and the southeastern Ukraine of Soviet Russia). Whether and how far man, through grazing and plowing, has accelerated natural rates of wind erosion of these dust soils is still a controversial point. Regardless of the outcome of this controversy, it is relevant for public policy to ask whether it is economical to influence the effects of wind upon soil movement, soil moisture, and directly upon plants. The experience with shelter belts and several other conservation practices gives an affirmative answer to this question—at least for some areas.

In flood and stream control, likewise, conservation policy deals largely with natural phenomena. Here, also, a controversy rages how far the latter have been aggravated by man—for example, through clear cutting and repeated burning of forests. There is some indication that these and other man-made changes of watersheds are more important for reducing the quality of runoff (increased load of gravel and debris, increased water temperature, lowered oxygen content, and so on), than for increasing flood peaks. Still, it may be economical to modify both natural and man-made conditions of a watershed—through reservoirs, channel rectification, fire protection, land-use practices, and so on.

Although it is obvious, it may not be entirely amiss to mention that, in order to decide whether or not it is economical to modify natural phenomena, their beneficial effects must be as fully considered as their damages. Geologic erosion by wind or water and floods frequently have such beneficial effects (p. 256).

PART V

THE SOCIAL ECONOMICS OF CONSERVATION

IMPLEMENTATION OF CONSERVATION POLICY

CHAPTER

19

DOMESTIC TOOLS OF CONSERVATION POLICY

Objectives Only One Aspect of Policy

Clarification of the problems involved in determining objectives of conservation policy covers only a part of the social economics of conservation. Equally important are the issues involved in achieving these objectives in actual life. These issues can be divided into two parts.

There is first the problem of selecting the proper tools. This problem will be discussed for domestic policies in the current chapter. The international aspects of such tools will be taken up in the next chapter. It will appear that there are many tools, domestic as well as international, which can be used simultaneously.

The complementary and competitive relations between tools pose the problem of their coördination in both the legislative and executive branches of governments at various levels. This problem will be taken up in chapter 21.

The objectives and criteria of conservation policy were intentionally treated on a rather general level in order to make our results meaningful under various institutional conditions. When it comes to the implementation of conservation policy, this level of presentation cannot be continued. In this part of our study

(Part V), the focus is on implementing conservation policy by governments at various levels *in the United States.*[1]

Direct and Indirect Tools of Conservation Policy

Among the many tools of conservation policy, two main groups—direct and indirect ones—were differentiated in chapter 16. The most important members of the latter group have already been reviewed. These are the economic forces and social institutions discussed in chapters 7 to 15. The many ways in which these forces and institutions can be tools as well as obstacles of conservation policy were pointed out. Such tools may be called "indirect" because changes of interest rates, of uncertainty, of prices, of taxation, of tenure, of monopolistic condition, and the like are usually brought about by governments with objectives other than changing the state of conservation. But, as was pointed out before, even if conservation objectives were taken into account, the actual results of such policies in terms of conservation or depletion would be uncertain in degree, geographical extent, and timing.

In contrast, direct tools of conservation policy are employed largely in the interest of conservation objectives. Their results in terms of conservation and depletion are fairly certain with respect to degree, geographical extent, and timing. Such direct tools vary in type from government-subsidized education to strict zoning ordinances and the requirement (or prohibition) of specific conservation (depletion) practices.

Indirect tools rely for their effectiveness on the economic calculations of individual planning agents and on the process of economic selection. The effectiveness of some direct tools—education, subsidies, and penalties—depends on the same factors. However, most direct tools rely, in addition, on the police power to bring about changes in the state of conservation in private enterprise.

Such reliance is necessary in many practical situations: If economic inducements alone are operative, there is little probability

[1] In the Spanish edition of this study, an additional Part VI deals with the implementation of conservation policy *in Latin American countries.*

that all planning agents will respond to a certain uniform degree in a certain area within a certain period of time. Only *after* economic inducements have been tried is it possible to ascertain the *aggregate* effects and to utilize this experience for forecasting probable effects in similar future situations. However, the success of conservation policy frequently requires that all planning agents in a certain area respond to a certain degree within a certain period of time. A few examples may illustrate this point.

If the scenic beauty of a landscape or the character of a residential district is to be protected, nonconforming uses must be excluded without exception; economic incentives alone would not guarantee this exclusion. Similarly, if only a certain number of game animals are to be harvested each year, it may be necessary to limit the take per hunter and the number of hunters; a per-capita tax (per hunter or per animal taken) would be too uncertain a tool of conservation policy. To prevent wind erosion in a certain critical area, success may depend on the coöperation of a few absentee landowners whose economic interest in the welfare of the local community may be so small that an educational appeal would be ineffective and the necessary subsidies too costly and uncertain. In such cases land-use regulations by a soil-conservation district formed under democratic procedures, equipped with taxing and regulating powers, and backed by the machinery of the courts may be the most efficient and equitable solution.

Among the direct tools, economic problems of conservation subsidies or penalties were taken up earlier with the effects of income, uncertainty, prices, and taxation. Thus, there are left at this point mainly the tools of education, zoning, and regulation of practices. These will be taken up in turn. For reasons explained in the preceding chapter, emphasis will be placed on the problem of guarding against economic irreversibility of depletion—that is, on realizing a safe minimum standard of conservation for critical-zone resources.

Education

Most workers in the field of conservation emphasize the importance of education. The need for conservation education is uncontroversial—in contrast to some of the other tools of conservation policy discussed below. Rather than labor the obvious, it is more interesting to raise a few questions with respect to some generally accepted ideas about the role of conservation education. These questions concern: who is to be educated, in what subjects, by whom, and with what probable success?

Education of the Voting Public

When the importance of conservation education is stressed, most people think about the ignorance of resource users. Admittedly, the knowledge of resource users is frequently imperfect in the sense that they do not employ the best knowledge available in a given social group. Types of such imperfect knowledge will be discussed presently.

Overcoming such imperfections is only one aspect of the conservation education that is needed. The preceding chapters tried to show that resources are depleted not so much because resource users do not know any better, as because they cannot help it under the influence of economic forces and social institutions. To change these forces and institutions in a democracy requires the awareness of the voting public.

The voting strength in the United States during the last half century has shifted heavily towards those segments of the population who do not use resources directly in agriculture, forestry, grazing, mining, and other extractive industries. Moreover, a greater and greater proportion of the voters reside in urban communities where their direct contact with problems of resource use is entirely absent or confined to occasional recreational use. These shifts in occupation and residence are still in progress. Thus, awareness of resource problems among the urban population, among industrial labor, management, trading, banking, and the professions appears as an important prerequisite of conservation policy.

Education of Resource Users

Turning now to the imperfect knowledge (in the sense explained in chapter 4) of resource users, emphasis in education is usually placed on the technology of conservation practices. The technological aspects of stopping gullies, sheet erosion, overgrazing, water pollution, damage to forest regeneration, and so on are certainly important. By and large, however, these aspects are fairly simple and easily understood by resource users—at least, if they are aware in time of the existing state of conservation and of the direction and rate of its change. However, such awareness is frequently absent. In other words, imperfect knowledge often exists with respect to a proper diagnosis rather than with respect to the remedies of depletion.

Sheet erosion, the first and (from the standpoint of reversibility) decisive phases of gullies, depletion of organic matter and other important components of soil productivity, degeneration of plant associations of rangelands and forests, impairment of natural forest regeneration, pollution of streams, secular (in contrast to cyclical) overdraft of ground water, and many other indications of depletion are not easily noticeable at once. The danger of their economic irreversibility is impressed upon individual planning agents only after some long and costly delay. Thus, emphasis in conservation education may well be shifted from the technology—frequently only the "mechanics"—of remedies to the requirements of an early diagnosis. Education may well be directed toward an understanding of changes in plant indicators, toward a sharp continuous observation of soil changes under different plant cover and treatment, and toward an ecological thinking, especially toward an appreciation of the relation existing between changes in the ecological balance and the danger of economic irreversibility.

Difficulties of Early Diagnosis

The difficulties of an early diagnosis of depletion and of the danger of its economic irreversibility are increased if depletion of critical-zone resources is obscured for individual planning agents and, in the aggregate, for a whole country, because other factors which increase revenues or decrease costs operate at the same time.

Soil depletion in the United States may be mentioned as an illustration: Plant varieties with higher yield per acre and greater resistance to diseases came gradually into use during the last three quarters of a century. Similar improvements took place in the efficiency of livestock to transform feed into meat, wool, and milk. Until 1920, prices of agricultural products increased both absolutely and in relation to the prices of other commodities, as the economic distance between farm and market was gradually reduced through increase in population and improvement of transportation by water, railroads, and highways. Costs per acre and per unit of product decreased for the same reasons and through improvements and inventions in the field of agricultural technology. Decreases in yield per acre due to soil depletion did not find expression in official statistics, because worn-out acres were replaced by newly occupied ones as long as the supply of virgin land lasted.

Only in the two decades after 1920, when some of these factors ceased to operate, did the individual farmer and the nation begin to realize the actual state of soil depletion. The last extensive tracts of virgin soil had been occupied during the First World War. The increase in population was slowed by the falling birth rate and by restriction of immigration. Foreign agricultural markets were closed partly for noneconomic reasons, partly because of the economic depression of the early 1930's. In bulk transportation there were no revolutionary changes comparable with the advent of canals, railways, ocean steamers, trucks, and improved highways. It is no accident that the conservation policy connected with the name of Franklin D. Roosevelt started during this period. In contrast, the conservation movement linked with the name of Theodore Roosevelt had been concerned more with the forests than with the agricultural resources of the country, because the state of forest depletion was more spectacular in his day.

Today, high prices resulting from the Second World War and its aftermath and rapid improvements in agricultural technology (hybrid corn, better varieties of plants and animals, more and better fertilizers, mechanization of cultivation and harvesting) tend to dull the awareness of farmers and governments of the actual state of soil conservation and its change.

Similar illustrations could be taken from other resources—for example, in the depletion of ground water.

Education of Different Age Classes

As in all educational effort, an attempt should be made to catch them young. Problems of resources and of human ecology would seem an excellent focus to teach citizenship in grade and high schools. On behalf of teacher training, these problems should receive more attention in colleges and universities.

The educational effort by governments should be directed especially toward those sections of the population where educational opportunities are inferior. Economic and occupational factors are frequently responsible for differences in educational opportunity. An example is agriculture. Although in the United States, as elsewhere, the educational level is lower in rural than in urban areas, the economic burden of education is far greater in country districts.[2] On the other hand, conservation education is especially needed in rural schools. In the United States conservation in agriculture has become economical for private enterprises only recently. Conservation, therefore, is not instilled in a young farmer through the example of his elders or his neighbors, as in most European countries. In the United States education must substitute for tradition and custom.

Emphasis on conservation education in schools does not mean

[2] In 1930 the farm population of the United States received 9.03 per cent of the national income, but had 30.63 per cent of the children from five to seventeen years old. This discrepancy in the percentage of national income and children of school age is greatest in regions with low farm incomes. For instance, the farm population in the eleven southeastern states received 2.21 per cent of the national income, but had 13.42 per cent of the nation's children of school age. Cf. U. S. National Resources Committee, *The Problems of a Changing Population* (Washington: Govt. Print. Off., 1938), Section 8, pp. 193–221, particularly p. 207.

For a discussion of differences in intellectual development of school children from different occupational classes, see Frank Lorimer and Frederick Osborn, *Dynamics of Population; Social and Biological Significance of Changing Birth Rates in the United States* (New York: The Macmillan Co., 1934), pp. 158–164.

that the adults should be written off as hopeless. Periodicals, films, radio, and television provide effective media to reach adults. Several private organizations and foundations—for example, Friends of the Land and The Conservation Foundation—are actively engaged in this field.

As already suggested, there is one type of education which is mainly concerned with adults—that is, the voters; for in a democracy, education is essential for legislating and executing conservation policy. Often, conservation policy cannot rely on voluntary coöperation alone. An understanding of the need for and the objectives of compulsion is necessary for the enactment and enforcement of laws and regulations discussed below. An example is the successful educational campaign of the state of Wisconsin in making rural zoning acceptable.

Educational Agencies

To what government agencies should conservation education be entrusted? No major problem exists in the education of youth. Improvements can be made by better training of teachers, but not by adding special conservation agencies to the existing educational establishment.[3]

In adult education, a number of government agencies are active in conservation education. An example is agriculture, where the Agricultural Extension Service (which also is responsible for 4-H Club work),[4] the Soil Conservation Service, the Agricultural Conservation Programs Branch of the Production and Marketing

[3] In some states—for example, California—the state department of education is well aware of its opportunities in the conservation field and coöperates with the technical agencies of the state, such as the department of natural resources. The office of the chief of conservation education is established in the latter department. The two departments, together with the California Conservation Council—which also focuses on education in schools—recently sponsored regional conferences for teachers on conservation of natural resources (Pasadena, California, February, 1950, and Berkeley, California, January, 1951).

[4] The 4-H clubs are a great potential force in conservation education of youth. Many soil-conservation projects are undertaken by members. For the first time in the history of 4-H Club work, national awards for achievement in soil conservation were made at the 23d National 4-H Club Congress (Chicago, December 3–6, 1944). This is now an integral part of their program. Cf. National Committee on Boys and Girls Club Work, *National 4-H Awards Handbook—1951* (Chicago, Ill., 1951), pp. 58–61.

Administration (previously Agricultural Adjustment Administration), and other agencies of more regional importance [5] are involved in educational activities. The need for coördination of these agencies will be discussed in chapter 21. At this point, a few words about their activities in the field of conservation education must suffice.

The Agricultural Extension Service has only lately displayed a major interest in problems of soil conservation. Many workers in this agency were trained during a period when soil depletion was economical for private enterprise and when immediate expansion of production and effective marketing of products seemed the most pressing need. In addition, funds for developing the new field on a scale commensurate with its importance have not been available. In conservation education, two approaches, through demonstration farms and through assistance to individual farm planning, are most effective. Greater use of these approaches depends on personnel and funds.

Beginning in 1933 the Soil Conservation Service used a large part of its considerable funds for education through farm demonstration and assistance within the framework of soil-conservation districts. More will be said about the organization of these districts in connection with land-use regulations. The Soil Conservation Service was established when, for reasons discussed above, farmers were awakening to the actual state of affairs in soil depletion. The Soil Conservation Service has done much to hasten the awakening.

Opposition of some Extension Service workers to soil-conservation districts and rivalry between the services in some states have occurred. Coördination of government activities in soil-conservation education through information (Agricultural Extension Service) and through business service (Soil Conservation Service) is needed. At present, the Extension Service is not sufficiently well equipped if it comes to applying conservation education farm by farm, field by field, and month by month. Such

[5] For example, the Tennessee Valley Authority, the Grazing Service (now in the Bureau of Land Management, Department of the Interior), and the Forest Service of the Department of Agriculture.

help is best rendered by technicians employed by soil-conservation districts, who may be paid entirely, or partly, by public funds. How such service may be integrated with the present functions of the Agricultural Extension Service will be discussed later.

The Agricultural Adjustment Administration, during the second phase of its existence, from 1936 onward, effected education in soil conservation.[6] The subsidies offered for soil-conserving crops and soil-conserving practices,[7] the penalties imposed upon soil-depleting crops, and the formation of county planning committees have made farmers more conscious of the problems. But the definitions of soil-conserving crops and soil-conserving practices were determined to a considerable extent by considerations of production control that would obtain parity prices. In some instances, bona fide conservation became compromised with these and other objectives which were not necessarily in agreement with it. Here also coördination is needed with the activities of the Agricultural Extension Service and the Soil Conservation Service.

One other organization that may be mentioned at this point is the Civilian Conservation Corps.[8] Although its main objective was to mitigate unemployment among young men, its educational

[6] The Agricultural Adjustment Act of May 12, 1933 [48 *U. S. Stat. at L.* (1933), 31] was invalidated January 6, 1936, through decision of the Supreme Court in *United States v. Butler* (1936), 297 U. S. 1; 56 Sup. Ct. 312; 80 L.ed. 477.

Its successor, the Soil Conservation and Domestic Allotment Act of February 29, 1936 [49 *U. S. Stat. at L.* (1936), 1151], appeared as an amendment of the Soil Erosion Control Act of April 27, 1935 [49 *U. S. Stat. at L.* (1935), 163]—commonly called the Soil Conservation and Domestic Allotment Act—and, ostensibly at least, placed less emphasis on curtailment of production and on parity prices and more emphasis on soil conservation. At present, the soil-conservation activities of the A.A.A. are continued by the Agricultural Conservation Programs Branch of the Production and Marketing Administration.

[7] These subsidies are at present by far the largest single item—nearly 90 per cent—of all federal appropriations for soil conservation.

[8] The Civilian Conservation Corps was established by executive order April 5, 1933. Only three months after its establishment, 300,000 youths were enrolled in about 1,500 camps. In September of 1935, enrollment reached a peak of 505,782, and the camps numbered 2,652. From 1939 onward, enrollment decreased with the improvement of employment under the impetus of war preparations. Cf. Robert Fechner, *Objectives and Results of the Civilian Conservation Corps Program* (Washington: Civilian Conservation Corps, 1938), 35 pp. Processed.

achievements in the field of conservation should not be underrated. Nearly all CCC boys were employed in conservation projects. They not only became aware of the problems, but learned the technical means to solve them—building of terraces, stopping of gullies, use of vegetation to heal or prevent erosion, treatment of runoff. This educational by-product should not be forgotten if reëstablishment of a similar organization should ever be considered.

There is one educational approach which in this country has been tried for academic and government personnel but seldom if ever for resource users: Apprenticeship and in-training schemes may be established whereby young farmers, livestockmen, forestry workers, and the like, may be trained for a limited period in well-managed private enterprises and paid by these enterprises according to labor productivity. Such wages may be supplemented by government payments as needed to give a certain prestige (fellowship idea) to the work, as done for veteran trainees after the Second World War.

Education Is Not a Cure-All

In conclusion, there are certainly opportunities for conservation education in the United States. On the other hand, the possible results from education alone are frequently overestimated by many conservation enthusiasts. It was suggested above that the reason for such too high expectation must be sought in an underestimation of the strength and complexity of the economic and social forces which impinge upon resource users. The fact that resource users do not behave as they "should" from the standpoint of some observers is so easily "explained" by ignorance. Education, therefore, seems such a logical and simple solution.

Unfortunately, the problem is more difficult than that. If economic and institutional factors stand in the way, results from education alone will be small. Such an experience may discourage an educational approach even under conditions where it is well suited. Education should be used as a tool of conservation policy where it can be effective. Sometimes it is effective only if used in combination with other tools, such as zoning and regulation of practices. These tools will be taken up next.

Zoning

Zoning, at present, is the most common direct tool of public land-use policy. Usually, however, it is employed for objectives other than the conservation of natural resources. Often public health and safety are the immediate objectives. Zoning for height, area, and vision clearance in cities is not a tool for conserving site value; likewise, flood-plain zoning is usually not aimed at water or soil conservation. On the other hand, zoning of residential districts against stores and against many other nonresidential uses is concerned, not with public health and safety, but with conserving a certain type of use.

Zoning can be successfully applied in the conservation of several natural resources: Soil resources which, because of topography or semiarid conditions, have high erosion hazards and are suitable permanently only for grazing, may be zoned against other agricultural uses. Important infiltration areas for ground water may be zoned against urban development; as a joint product flood-control problems can thereby be solved in a way which is sometimes the most economical one. Forest resources may be protected by zoning against agriculture, grazing, or year-round residence. Sometimes all uses except one are excluded—for example, in watersheds near important population centers with semiarid climate (southern California). In the case of scenic resources, degrees of intensity in zoning may vary from billboard zoning to complete prohibition of all uses except temporary visits in wilderness areas.

Economic Characteristics of Zoning

Zoning has certain characteristics which under some conditions make it the most economical tool of conservation policy. Other characteristics impose economic limits on zoning.

The spatial character of zoning renders it well suited for reducing undesirable effects upon the state of conservation which originate through what was called "neighborhood relations" in resource utilization. Such relations are the effects which utilization of one piece of property has upon the utilization of another piece of property because of the location of the two properties. Sometimes these effects are based on highly complex cultural conditions as, for example, the effects of the race or occupation of the user of one property upon the possible uses of an adjoining property; or the burden to local communities—in terms of taxes for roads, schools, police—of a few scattered settlers in outlying areas. In other cases these effects are fairly simple; for example, they may be based on the actual spatial movement of water, fire, fumes, insects, and pests from one property to another.

Economically speaking, these neighborhood relations are problems of allocation; this means that some users do not bear all the costs or do not realize all the revenues which are functionally related to the use of their resources. This problem of allocation was encountered in connection with tenure. Its broad social significance for conservation policy was discussed in chapter 17. At this point we may note merely that zoning is an effective tool of public policy when problems of allocation are based on the spatial relations between users of resources.

If spatial relations are not in the picture, zoning is not an effective tool to prevent undesirable uses and allocation. For example, rare species of animals or plants are protected for an individual state or for the nation by prohibiting their killing, capture, or destruction, rather than by zoning against hunting or other uses.

Zoning is concerned with preventing uses, not with changing practices. Zoning is mainly negative; it prohibits nonconforming uses but does not require positive actions. Direct public controls that regulate practices are generally not applied as zoning ordinances. Such regulation will be considered separately later.

Economic and Legal Limits of Zoning

Zoning decreases economic opportunities for some uses and individuals and increases them for other uses and individuals; usually, only decreases are considered in the literature on zoning. Such decreases and increases occur mainly in the area zoned, but the economic repercussions are not necessarily confined to that area—for example, zoning of submarginal land against permanent settlement may result in tax savings for a whole county or state.

How far an interference with private utilization plans is constitutionally possible ("reasonable" and "not arbitrary") through use of the police power, the basis of zoning, has occupied lawyers, administrators, and courts during the last twenty years. All zoning is based on state enabling laws granting the power of zoning ordinances to local governments and in some cases to state agencies themselves. Such laws and ordinances are subject to the limitations of federal and state constitutions. In the United States, the legal limits of zoning have been expanded by the courts constantly as accepted social standards have changed with respect to the degree of public interference with private enterprise. Some states are backward with regard to zoning. Generally, however, in the United States, zoning has and probably will have sufficient legal scope.[9]

In the United States, the legal limits of zoning, as interpreted by the courts, are closely connected with the economic limits. The latter are of special interest in our study.

With severe zoning ordinances, all uses become uneconomical under private ownership. Many rural zoning ordinances, therefore, permit continuation of nonconforming uses for present owners in order to avoid unreasonable hardships, which might endanger the constitutionality of the measure. Areas zoned against most uses for the protection of water resources near big cities must generally be owned by public-utility districts. This type of public control involves the exercise of eminent domain and, commonly, full and final compensation for all existing use rights.

[9] Summaries of the extensive legal literature on zoning are found in:

Ralph B. Wertheimer, "Constitutionality of Rural Zoning," *California Law Review*, Vol. 26, No. 2, January 1938, pp. 175–205.

George A. Warp, "The Legal Status of Rural Zoning," *Illinois Law Review*, Vol. 36, 1941–1942, pp. 153–170.

Edward M. Basset, *Zoning; the Laws, Administration, and Court Decisions during the First Twenty Years* (New York: Russell Sage Foundation, 1940), 275 pp.

Compensation in Connection with Zoning

One may consider expanding the economic limits of zoning by means of compensation (subsidies). Zoning as now practiced does not involve compensation because private use rights are always enjoyed under an implied public limitation and must yield to the police power. In other cases of direct public controls, partial compensation is provided by law, such as in the regulations designed to eradicate bovine tuberculosis.

If compensation is given for a redistribution of use rights—direct public controls, essentially, involve such a redistribution—there will arise the problems of partial compensation to individual strands of the bundle of property rights and of variations of compensation over time with changing economic conditions. Both problems are economically and administratively rather difficult to solve. If the owner feels that use rights taken away from him have become more valuable, he will probably claim an upward revision of his compensation. If such claims are advanced by strong political pressure groups, they may be successful even if the original compensation was agreed upon as final. On the other hand, changing economic conditions may impel the public to exclude different use rights at different times. If, therefore, the economic limit of zoning without compensation is approached, the expansion of public ownership rather than of zoning plus compensation may be more economical.[10] Conceivably, this may hold true even if a fair and full compensation (for all use rights before zoning) is paid to owners whose properties are taken over

[10] Differences between police power (zoning) and eminent domain are differences of degree. This is well stated by Justice Oliver Wendell Holmes: "Let me suggest as to the Police Power that the phrase is used by the Courts in two senses—one, the general powers of legislation in which sense it is worse than useless: the other, as what I have called a conciliatory phrase—to cover those cases where the legislature is held justified in contravening the literal meaning of constitutional protections and is somewhat, for instance, limiting (*i.e.* taking) property rights without compensation. It is a matter of degree and in *Martin v. District of Columbia,* 205 U. S. 135, 139, I took pleasure in pointing out that a man's constitutional rights, the difference between the police power and the need of eminent domain with compensation, might be a matter of feet and inches. An interesting case when Brandeis and I differed arose last term *Pennsylvania Coal Co. v. Mahon,* Dec. 11, 1922, not yet in regular reports."—Harry C. Shriver, *Justice Oliver Wendell Holmes* (New York: Central Book Co., 1936), pp. 168–169. (See also pp. 299–300 and footnote 49.)

by the public. Not being concerned here with income redistribution, we may assume that such a compensation will be paid. If this assumption is granted, zoning may be used as an inducement to sell voluntarily to the government.

Zoning and Eminent Domain

Under our assumption regarding compensation, exercise of eminent domain contains fewer elements of confiscation than many zoning ordinances. At present, public opinion and, accordingly, the courts, impose severe limits (severe as compared with such measures in other countries, for example, in Great Britain) upon the exercise of eminent domain.[11] Because of these limits, governments usually find it more advisable—economically and politically—to purchase private property on a voluntary basis than to resort to eminent-domain proceedings.

In conservation policy, zoning and eminent-domain proceedings are not competitive, but complementary. Factors of economic location often operate so that in the heart of a problem area economic limits exclude zoning, whereas on the fringes zoning is economically possible. Zoning as well as eminent-domain proceedings may be supplemented by measures directed toward transferring resource users to other localities or occupations if such movements become necessary: compensation may be paid in the form of assistance to relocation.

---

[11] A recent exception is the use of eminent domain in the redevelopment of blighted areas in cities. In California, for example, the Community Redevelopment Act [*Cal. Stats.* (1945), pp. 2478–2500] enables any city or county to require the sale to the public of land and buildings in any blighted area. It permits the city or county then to clear the land of the structures upon it, and to sell or lease the area to any group of businessmen or private developers, provided they agree to rebuild the area for industry or residences in accordance with a general plan laid down in advance by the city or county.

The law safeguards the rights of present property owners by carefully defining blighted areas and specifying that redevelopment may be undertaken only if those areas threaten the health, safety, and general welfare of the people of the communities in which they exist and of the people of the state generally. It provides that present owners have prior right to participate in the redevelopment plan, if they wish, and that they shall be fully compensated for their property if they do not wish to participate.

Regulation of Practices

Direct public regulation of practices is applied not only through ordinances by counties or cities—as is zoning—but also by state or federal statute and by special districts. In resource utilization, such special districts are of particular importance. Examples are irrigation and water districts, districts for soil conservation, forest conservation, grazing, fire control, and placer mining.

Such districts, which are regulating authorities, should be clearly differentiated from zoning districts, which are merely a spatial form of control. The former districts are established and their operations regulated by enabling statutes, both federal and state. Generally, their use as tools of public control is safeguarded by the democratic processes of referendum and election.

Of the many types of regulation, a few concerning the most important resources—water, soil, forests, grasslands, wildlife, oil, gas, and minerals—may be considered.

Regulation of Water Use

Among the most important privately used resources (that is, excluding some resources of smaller economic importance, such as wildlife, and some publicly managed resources, such as helium) public regulation of practices has gone furthest with water—both in social philosophy and in the development of institutional mechanisms for control. This is particularly true for the semiarid states. Public regulation of water use in California may serve as an illustration.

The importance for conservation of rights and customs of resource tenure was emphasized. At this point we are not so much concerned with water rights—for example, with riparian versus appropriation and prescription rights or with the doctrine of correlative rights—but with public regulation of water use. However, the problems of water rights cannot be entirely separated from the problem of regulation. In California, the principle of "reasonable, beneficial use," which is the basis for regulation, was developed in the long course of the economic and legal conflicts between riparian and appropriation rights. The principle was embodied in the statutes regulating appropriation of water as early as 1872,[12] was extended to ground water by the courts in 1903,[13] and has been anchored in the California constitution since 1928.[14]

What constitutes "reasonable, beneficial use" is defined in terms of practices through a number of institutional mechanisms, namely, (1) through statutes; (2) through rules of the California Division of Water Resources (which administers water appropriation); (3) through court decisions; (4) through county ordinances; and (5) through regulations of irrigation (water) districts. A few examples of these definitions may be given.

Through statute, "reasonable, beneficial use" is defined for artesian waters [15] and for the irrigation of uncultivated land.[16] Likewise, through statute, spreading of surface water for replenishing ground water—which for some time gave rise to litigation—is specifically designated as "reasonable and beneficial." [17]

The rules of the Division of Water Resources define in some

---

[12] Section 1411 of the California Civil Code of 1872 states: "The appropriation must be for some useful or beneficial purpose, and when the appropriator or his successor in interest ceases to use it for such a purpose, the right ceases."

[13] *Katz v. Walkinshaw*, 141 Cal., 116; 70 Pac., 663 (1902); 74 Pac., 766 (1903).

[14] Because of its clear statement of a basic social philosophy, the relevant passage of Article XIV, section 3 of the California constitution may be quoted: "It is hereby declared that because of the conditions prevailing in this state the general welfare requires that the water resources of the State be put to beneficial use to the fullest extent of which they are capable, and that the waste or unreasonable use or unreasonable method of use of water be prevented, and that the conservation of such waters is to be exercised with a view to the reasonable and beneficial use thereof in the interest of the people and for the public welfare."

[15] The basic act is that given in *Cal. Stats.* (1907), pp. 122–123.

Artesian wells can be used only for beneficial purposes, viz., for domestic use, irrigation, and the propagation of fish. If more than 5 per cent (10 per cent in the basic act) of the artesian supply escapes from the land on which it is developed, this constitutes a punishable offense. The act is administered by the Division of Water Resources.—*Cal. Water Code* (1949), secs. 300–311, pp. 24–25.

[16] The quantity of water which may be used for the irrigation of uncultivated areas of land not devoted to cultivated crops is limited to not more than 2½ acre-feet per acre in any one year. This is interpreted to mean net use on the land exclusive of reasonable transportation losses, which losses may be included in the amount of water applied for.—*Cal. Water Code* (1949), sec. 1004, p. 26.

[17] *Cal. Stats.* (1919), chap. 423, pp. 826–827.

detail the quantities of water which can be regarded as "reasonable, beneficial use" for different domestic purposes, and for different conditions of irrigation. Authority for these rules rests in the statutes.[18] The rules themselves are easily accessible and need not be given here.[19] Similar rules will probably govern the administration of the legislation recently enacted to combat water pollution.[20]

The courts have adopted the position that there can be no general quantitative definition of "reasonable, beneficial use." Definition of this principle depends on the facts in each case. Nevertheless, a number of practices have been declared in conflict with the principle. Examples are winter irrigation to get rid of gophers,[21] use of water for duck clubs,[22] sinking a well near an irrigation ditch,[23] and lowering the water level in wells of adjoining landowners, so that they are put to considerable expense for deepening their wells.[24]

Statutes, the courts, and the Division of Water Resources cooperate in the detailed allocation of water use—setting forth amount, season of use, purpose of use, point of diversion, and priorities. The two procedures are the court reference procedure and the statutory adjudication procedure. The first is applicable to all waters.[25] It merely permits the court to refer any case to the Division of Water Resources for investigation and information of the court prior to its decision. The second procedure is applicable only to surface water and ground water in definite

---

[18] *Cal. Water Code* (1943), secs. 1050–1058; *Cal. Stats.* (1943), chap. 368, pp. 1611–1612.

[19] State of California, Division of Water Resources, *Rules, Regulations and Information Pertaining to the Appropriation of Water in California* (Pamphlets issued in Sacramento: 1938, 1941, 1948, and 1949).

[20] The so-called Dickey Bill enacted by the California Legislature in August, 1949. —*Cal. Stats.* (1949), chap. 1549, pp. 2782–2789.

See also *Report of the Interim Fact-Finding Committee on Water Pollution.* (Sacramento: Assembly of the State of California, 1949), 172 pp.

[21] *Tulare Irr. Dist. v. Lindsay-Strathmore Irr. Dist.*, 3 Cal.2d, 489; 45 Pac.2d, 972, (1935).

[22] *Ex parte Elam*, 6 Cal. App., 233; 91 Pac., 811 (1907).

[23] *Lemm v. Rutherford*, 76 Cal. App., 455; 245 Pac., 225 (1926).

[24] *City of Lodi v. East Bay Municipal Utility Dist.*, 7 Cal.2d, 316; 60 Pac.2d, 439 (1936).

[25] *Cal. Water Code* (1943), secs. 2000–2076; *Cal. Stats.* (1943), chap. 368, pp. 1630–1632.

channels.[26] Under it any claimants may file petitions with the Division of Water Resources for investigation. After the investigation, the division files its allocation with the Superior Court. The Superior Court hears protests and hands down a ruling. In both procedures, provision is made for the administration of court decisions through a Water Master appointed by the Division of Water Resources.

At least one California county (Orange County) has adopted an ordinance prohibiting use of water contrary to the principle of "reasonable, beneficial use." The ordinance is patterned after the act preventing waste from artesian wells mentioned above.[27]

California irrigation districts normally include among their rules and regulations a provision for the prevention of unreasonable use. The following regulation is typical: [28] "Persons wasting water on roads and vacant land, or land previously irrigated, either willfully, carelessly, or on account of defective ditches or inadequately prepared land, or who shall flood certain portions of the land to an unreasonable depth or amount in order to properly irrigate other portions, will be refused the use of water until such conditions are remedied."

Regulation of Agricultural Use of the Soil

The socially most important resource, the soil used through agriculture, is affected by fewer regulations of practices than one might expect.

In principle, the common-law approach could be applied if accelerated erosion caused by agricultural practices on one property damages another. But this approach is costly, slow, uncertain, and in practice is rarely if ever tried.

---

[26] *Cal. Stats.* (1943), chap. 368, pp. 1632–1642.

[27] Waste is defined as letting water flow from a well into a stream channel, pond, bay, etc., unless later used for a beneficial purpose (domestic use, irrigation, or propagation of fish), or permitting more than 5 per cent of the water pumped to waste off the land on which it was used.

This ordinance was upheld by the court in *In re Maas* (1933), 219 Cal., 422; 27 Pac.2d, 373, on the grounds that under Article XI of the constitution the county has power to make police regulations which do not conflict with general laws.

[28] S. T. Harding, *Operation and Maintenance of Irrigation Systems* (New York: McGraw-Hill Book Co., Inc., 1917), p. 183.

Likewise, the public-nuisance principle, which is frequently employed to regulate practices in the use of other resources (oil, gas, and placer mining; see below) is of little practical importance in agriculture except as a basis for zoning ordinances.

The spectacular effects of wind erosion have led in a few states (Texas, Oklahoma, New Mexico, Oregon) to special statutes giving attention to land-use practices which affect this form of soil depletion.[29]

Soil-conservation districts have multiplied rapidly all over the United States since the 1930's through the efforts of the Soil Conservation Service in establishing and in assisting them.[30] Soil-conservation districts are formed through democratic procedures defined by state laws. Such soil-conservation-district laws, patterned after the federal "standard" law, are now enacted in all states.

At one time 33 of the 48 state laws granted regulatory powers to soil-conservation districts. Regulations based on such powers may require, for example, that steep or otherwise highly erosible land be retired from cultivation, that specific practices (contour cultivation, lister furrowing, strip cropping) be followed, and that structures (terraces, check dams, water outlets) be erected. Failure to observe the regulations may be punishable by fine, and the supervisors of the soil-conservation district may file a petition asking the county courts to order observance of the regulations. Such court orders may provide that if the land occupier fails to perform, the supervisors may go upon his land, do the necessary work, and collect the costs from him. A board of adjustment is provided in order to permit variances from regulations upon petition of the land occupier in cases where strict application would result in great practical difficulties or hardships. Decisions of the board are subject to review by the courts.

So far, the significance of soil-conservation districts for regulation of practices has been almost nil. Regulations have been

[29] *Tex. Stats.* (1935), chaps. 214 and 337; *Okla. Stats.* (1937), chaps. 250–254; *N. Mex. Stats.* (1937), chap. 222; *Ore. Stats.* (1937), chap. 131.

[30] There are more than 2,000 soil-conservation districts assisted by the Soil Conservation Service in 48 states, Puerto Rico, the Virgin Islands, Hawaii, and Alaska. These districts contain more than 1 billion acres.

adopted only by a few districts in three states (Colorado, North Dakota, and Oregon). Enforcement has not been tested before the courts. The wording of most state soil-conservation-district laws makes adoption of regulations rather difficult.[31] Recently, several states—among them California—have revised their laws in such a way that districts do not have regulatory powers. The revised California law prevents districts explicitly from doing any work on any land without the permission of the owner.[32] Although federal influence was largely responsible for granting regulatory powers to districts in state statutes, officials of the Soil Conservation Service have not been eager to get regulations actually adopted. Possibly they fear retarding the formation of districts, or the difficulties and expense for the districts which might result from testing in the courts the validity of regulations.

Only two state statutes (in California [33] and Colorado) give soil-conservation districts taxing power. Such power is the most important instrument for coöperative action in many other types of districts (for example, irrigation, flood control, fire control, and placer mining). Through lack of taxing power, soil-conservation districts become dependent upon the assistance of the Soil Conservation Service. On the other hand, it can be argued that granting the power to tax would have created the danger of inexpert financial management in a new field and would have discouraged the establishment of some districts. Furthermore, if regulatory and taxing powers are desired, laws in many states—for example, California—provide for county conservancy and flood-control districts [34] which have these powers.

The weakness of soil-conservation districts with respect to regulation of practices and independent financing does not imply that their influence on land use has been negligible. As stated earlier, their achievements have been largely in the field of education.

[31] A referendum must be held on the enactment of regulations. The favorable vote required ranges in some states up to 90 per cent of those voting. Appeal to the regular courts is always possible.

[32] Revision of Division 9, Public Resources Code enacted in July 1949.—*Cal. Stats.* (1949), chap. 1031, p. 1893.

[33] In California the taxing power is limited to 2 cents per $100 assessed value.

[34] According to California law, units of a county can be organized and taxation also may be within units.

So far, these achievements have been possible because of the assistance received through the Soil Conservation Service. However, soil-conservation districts could be an even more important institutional mechanism of coöperative self-improvement and self-regulation by the users of the soil.

Regulation of Forest Use

With respect to forest resources, the federal government regulates cutting practices and other private uses in federal forests in order to obtain proper regeneration, reduction of fire hazards, protection of watersheds, and preservation of scenic values. The federal government also coöperates with states in establishing and financing of fire districts and in insect and pest control.[35]

Most states have statutes providing for fire, insect, and pest-control regulations.[36] In addition, some states—for example, California and Oregon—have regulations aimed at proper restocking.[37]

An attempt is being made in California to regulate forest prac-

[35] Federal forest legislation is very extensive. The most important acts are: Act of 1897 [30 *U. S. Stat. at L.* (1897), 34]; Weeks Act of 1911 [36 *U. S. Stat. at L.* (1911), 961]; Clarke-McNary Act of 1924 [43 *U. S. Stat. at L.* (1924), 849]; Fulmer Act of 1935 [49 *U. S. Stat. at L.* (1935), 963]; Cooperative Farm Forestry Act of 1937 [50 *U. S. Stat. at L.* (1937), 188]; Act of 1944 [58 *U. S. Stat. at L.* (1944), 132].

[36] In California, for example: Fire-protection legislation of 1919 [*Cal. Stats.* (1919), pp. 262–264, and amendments, especially in 1923 and 1929]; Forest Insect Control Act of 1923 [*Cal. Stats.* (1923), pp. 156–158]. Furthermore, a recent revision of the California law empowers the board of supervisors to appropriate general tax funds if no district assessment is made (*Cal. Public Resources Code,* Div. 9, sec. 9301).

[37] A 1943 amendment to the California Forestry Conservation Law stipulates: "It is unlawful to cut for conversion into lumber any sound live coniferous tree that is less than 18 inches in average diameter measured outside of the bark at a point on the trunk six inches above the general level of the ground surrounding the tree, in that part of California lying north of the 6th Parallel, South, unless a permit so to do is first obtained from the State Forester as provided in this chapter."—*Cal. Stats.* (1943), pp. 1067–1068. (Amending *Cal. Public Resources Code,* Div. 4, chap. 7, to include sections 4850–4853.)

Section 4 of the Oregon Forest Conservation Act of 1941 provides that anyone cutting trees for commercial use must leave reserve trees of commercial species to maintain continuous forest growth or provide for satisfactory restocking. Operators must make every reasonable effort to protect such reserves from unnecessary damage resulting from logging operations or from fire while slash burning or at other times. —*Ore. Laws,* (1941), chap. 237, sec. 4, p. 372.

tices through a special type of forest district. The attempt by California to use the idea of self-regulation by districts in combination with statutory regulation and police power to improve forest practices appears of general interest for conservation policy. For this reason, a more detailed discussion seems justified.

The California Forest Practices Act of 1945 establishes four forest districts and a forest-practice committee for each district. These committees consist of five members, two of whom are private timber owner-operators, one a private timber owner owning more than 1,000 acres of commercial timber but who does not conduct logging or milling operations, and one a farmer who owns commercial timber of not less than 160 acres and not more than 1,000 acres. These four members are appointed by the governor. The fifth member is an official of the State Board of Forestry who acts as the secretary of the committee but does not have voting power. The district forest-practice committees adopt forest-conservation practices after due notice and hearings. When such practices have been approved by two-thirds of the private timber ownership of the district (to be measured by the acreage of timber *and* cutover land privately owned) they are submitted to the State Board of Forestry for approval. If approved by the board, the district forest practices have the force of law. The state forester is required to undertake periodic inspections to determine whether the rules are being complied with. All timber operators must register with the state forester. Any operator who fails to register will be prohibited from engaging in commercial timber activities.

All four districts are in operation. Each has drawn up and published a set of regulations. They were approved by the board and became law on September 4, 1947. The regulations are arranged under five heads—cutting practices, logging practices, fire-prevention practices, fire-suppression practices, and forest insect- and disease-protection practices.

As an example, the cutting, logging, and fire-prevention practices may be given in order to show the detailed nature of these regulations.

1. Cutting Practices

a. Redwood Forest District: In areas of old growth, 4 seed trees with a diameter not less than 24 inches are to be left after cutting operations. In areas of young growth, 8 seed trees with a diameter not less than 18 inches are to be left. In special-product areas (poles, piling, mine timbers, etc.) not more than one half the trees between 12 and 18 inches in diameter are to be cut in any ten-year period.

b. North Sierra Pine Forest District: Immature trees of less than 20 inches diameter at breast height (d.b.h.) are not to be cut. Special-product areas as in *a.*

c. South Sierra Pine Forest District: In areas of old growth west of the summit of the Sierra Nevada, immature trees of less than 22 inches d.b.h. are not to be cut, and 2 seed trees per acre are to be left. In areas of old growth east of the summit of the Sierra Nevada, the minimum is 18 inches d.b.h. Special product areas as in *a.*

d. Coast Range Pine and Fir Forest District: Immature trees with less than 20 inches diameter not to be cut. Two seed trees per acre to be left.

2. Logging Practices

To maintain the productivity of harvested lands, timber operators shall fell trees, when topography permits, in line with skidding direction and away from young growth, etc.; limit skid trails, skid roads, and "come-back" roads to the minimum width and number consistent with practicable and economic logging, and require tractor drivers to use them going to and from landings; require that guy lines and other rigging be hung on stumps or trees to be cut, but when necessary to use reserve trees; require that suitable lagging shall be used to prevent girdling; require that bunching, skidding, and return-trip tractor operations be performed without unnecessary damage to reproduction and immature and seed trees; instruct employees in good practices.

3. Fire-Prevention Practices

In slash disposal, burn all slash within 100 feet of public roads and main truck roads; do such burning away from residual trees and young growth; complete slash disposal prior to commencement of dry season; do all burning under proper weather and moisture conditions. In snag disposal, fell all snags 30 feet and over in height within 200 feet of public roads or main truck roads, and complete snag disposal prior to April 15 of the following year. Leave passable all truck and railroad grades at the end of each season for use in case of fire. Restrict smoking, lunch fires, welding, and blasting to safe areas and distances. Construct firebreaks not less than 8 feet wide around landings and motors between April 15 and December 1. (Such clearance is required in Section 4155 of the Public Resources Code.)

As yet, the State Board of Forestry has not attempted to enforce these regulations but has sought only coöperation. Thus there has been no court case to determine whether the board has the police power to enforce them. Some doubts have been expressed with respect to this power. There is also some difference of opinion with respect to the actual results obtained.

Regulation of Grazing Use

Grazing resources have been subject to some regulation of practices for a long time. The fence and stock laws of most states are examples; however, the primary objective of these regulations is not conservation. More recently a few states, for example, Texas,[38] have enacted grazing regulations with a view to preventing overgrazing and soil erosion; these regulations apply to seasonal periods of grazing and to density of stocking.

The most significant attempts to regulate grazing practices in the interest of conservation are being made by grazing districts. These districts were formed under the Taylor Grazing Act discussed in connection with the relations between tenure and conservation. Regulations of grazing practices apply to seasonal periods of grazing, density of stocking, drift fences, water improvements, and other aspects of use essential for conservation.

Regulation of Wildlife Use

Since wildlife resources in the United States are public property until captured, regulation of hunting and fishing practices rests on a secure and fairly simple institutional basis. Use of upland game and inland fisheries is under the jurisdiction of the states. Federal regulation is necessary if international aspects of wildlife conservation are involved.

State as well as federal laws contain detailed positive and negative requirements with respect to equipment (hunting arms, fishing gear), time of utilization during the day or the year, size and

[38] *Tex. Stats.* (1935), chap. 313.

kind of game and fish to be taken, information about yield which must be furnished to authorities, transportation and sale of yield, and many other matters. Details of these regulations are easily obtained by consulting the hunting and fishing codes available in every state.

Regulations of Oil and Gas Production

On the federal level, production of oil and gas is directly regulated through the Mineral Leasing Act of 1920 [39] and indirectly through the Connally Oil Transportation Act of 1935.[40]

There are many state statutes requiring or prohibiting practices. In California, for example, "unreasonable and willful waste" is prohibited. This waste includes blowing, release, or escape of gas into the air, production of gas in quantities exceeding a "reasonable" ratio to the amount of oil produced.[41] California law provides only for voluntary unitization of oil pools. Spacing of wells is obtained through application of the public-nuisance principle.

Several other states (Arkansas, Michigan, Texas, and several midwestern states) go much further in their oil- and gas-conservation laws. They provide for compulsory pooling, spacing of drilling, limitation of total production of each pool in the state, and prorating among individual wells.[42]

Regulation of Mineral Production

With respect to coal resources, the minimum-price regulations of the federal Bituminous Coal Conservation Act of 1935 [43] and its successor are of little interest for conservation—

[39] 41 *U. S. Stat. at L.* (1919–1921), 437, and amendments. Under the act, federal lands containing oil, oil shale, gas, phosphate, coal, and sulfur are leased to private users. Such leases generally make provision for practices intended to reduce "waste."

[40] 49 *U. S. Stat. at L.* (1935–1936), 30. The act aims at control of "contraband oil" —that is, oil transported over state lines after being produced contrary to direct state controls (production quotas) applied in connection with the Interstate Oil Compact of 1935.

[41] Oil and Gas Conservation Act of 1915 [*Cal. Stats.* (1915), chap. 718, and amendments]. See also *Cal. Public Resources Code* (1939), Div. 3, Oil and Gas.

[42] For example, Arkansas Oil and Gas Conservation Law of 1939 [*Ark. Stats.* (1939), chap. 105, sec. 14, pp. 230–232 and sec. 16, pp. 234–235].

[43] 49 *U. S. Stat. at L.* (1935–1936), 991, repealed and replaced by the Bituminous Coal Act of 1937.

in spite of the name of the act. Laws dealing with measures to prevent subsidence (pillar cutting) have importance for coal conservation.[44]

Interesting from the standpoint of competition between mining and agriculture is the regulation of strip mining in West Virginia requiring the operator to replace ". . . soil, subsoil or other strata removed from said coal and refill any ditches, trenches or evacuation made in stripping said coal." [45] This regulation applies also to the lands owned by the coal operator himself. It has been correctly observed [46] from the legal standpoint that the reasoning behind strip-mining regulations is identical with the Soil Conservation District Law of West Virginia, which was enacted in the same year.

In hydraulic mining, special federal and state legislation exists to prevent pollution of streams.[47] However, a bill to require dredging operators to replace the topsoil of agricultural land was defeated in the 1944 California Legislature. Some California counties are passing ordinances restricting dredging activity. Siskiyou County supervisors recently passed an ordinance requiring replacing 3 feet of topsoil on agricultural lands and leveling off the nonagricultural lands after dredging operations.

Utilization of phosphates and sulfur on federal land is regulated through the Mineral Leasing Act. In contrast, other minerals

---

[44] The best-known act is the Kohler Act of 1921 in Pennsylvania, which was declared unconstitutional by the Supreme Court one year later. [*Pennsylvania Coal Co. v. Mahon.* 260 U. S. 393; 43 Sup. Ct. 158; 67 L.ed. 322 (1922).]

See also C. C. Williams, Jr., "Conservation of Mineral Resources: A Brief Survey," *West Virginia Law Quarterly and the Bar,* Vol. 47, No. 4, June 1941, pp. 247–273.

[45] *W. Vir. Code* (1943), Art. 2A, sec. 2461(2); *W. Vir. Acts* (1939), chap. 84, pp. 402–403.

An adequate bond has to be furnished (with a minimum penalty of $150 per acre) before a permit for strip mining is issued.

[46] C. C. Williams, Jr., "Statutory Regulation of Strip Mining," *West Virginia Law Quarterly and the Bar,* Vol. 47, No. 1, Dec. 1940, pp. 52–57.

[47] Federal legislation is popularly known as the Caminetti Act [27 *U. S. Stat. at L.* (1893), 507, as amended by 30 *U. S. Stat. at L.* (1898), 631; 30 *U. S. Stat. at L.* (1899), 1148; 31 *U. S. Stat. at L.* (1900), 631; 34 *U. S. Stat. at L.* (1907), 1001; 48 *U. S. Stat. at L.* (1934), 1118; 52 *U. S. Stat. at L.* (1938), 1040].

Protection of fish from damage resulting from hydraulic mining operations during certain periods and in certain rivers is provided for by *Cal. Stats.* (1931), chap. 215, p. 387, as amended by *Cal. Stats.* (1937), chap. 306, p. 679; *Cal. Stats.* (1939), chap. 760, p. 2290.

on federal lands, chiefly metallic minerals, are still disposed of under the mining laws of 1866 and 1872.[48] It would seem desirable to revise these laws in order to provide for leasing under appropriate regulations, such as exist, for example, in some provinces of Canada (Alberta).

Production of one resource (helium) is entirely managed by the federal government; production of another (uranium ore) is dependent on government-determined prices and subsidies.

Regulation of the Processing of Resources

The common-law approach and the public-nuisance principle are frequently employed to regulate practices in the processing (in contrast to the production) of resources. Examples are regulations to abate air pollution caused by the fumes of copper smelters, and water pollution caused by waste products of pulp mills and oil refineries. However, regulation of the processing of resources is outside the immediate interest of this study.

Economic and Legal Limits of Regulation

Regulations, like zoning, pose the problem of the limits of direct public interference with private enterprise and, as a corollary, the problem of public ownership of resources. These limits are both legal and economic.

The legal limits themselves are essentially economic—that is, they depend on the degree of economic burden imposed upon private planning agents. This is well expressed by Justice Holmes in the decision of the Supreme Court declaring the Pennsylvania Kohler Act unconstitutional in 1922:

> Government hardly could go on if to some extent values incident to property could not be diminished without paying for every such change in the general law. As long recognized, some values are enjoyed under an implied limitation and must yield to the police power. But obviously the implied limitation must have its limits, or the contract and due process clauses are gone. One fact for consideration in determining such limits is the extent of the diminution. When it reaches a certain magnitude, in most if not in all cases there must be an exercise of eminent domain and

[48] 14 *U. S. Stat. at L.* (1866), 86, and 17 *U. S. Stat. at L.* (1872), 91.

compensation to sustain the act. So the question depends upon the particular facts. The greatest weight is given to the judgment of the legislature, but it always is open to interested parties to contend that the legislature has gone beyond its constitutional power.[49]

Since this decision was rendered, the courts have been conservative, but neither negative nor inflexible, in defining a "reasonable" degree of direct public control over private enterprise.

Compensation in Connection with Regulation

At present almost nothing is being done to expand the economic limits of resource regulation through compensation for the economic burden to private enterprises. Existing forms of compensation apply to practices that are voluntarily agreed on. They are, strictly speaking, subsidies paid with the aim to induce, not to compensate. Examples are the soil-conservation payments under the A.A.A. and a clause in the Taylor Grazing Act stating that preference in grazing rights on public lands is to be given for the proper use of private land and water.

It appears to be well within the scope of public policy to compensate for the costs of required practices so far as such practices do not benefit the resource user who is asked to perform them. The cost of practices and their benefit to resource users can be ascertained without great difficulty. The situation in this respect is, therefore, better than in zoning.

Compensation should always be given under conditions that would enable the government to reduce or to terminate it at any time, with or without terminating the requirement of practices. Technological and economic conditions may so change that in time the same practices may reasonably be required without compensation. In conservation policy this will happen frequently.

As mentioned earlier, compensation for public regulations

[49] *Pennsylvania Coal Co. v. Mahon,* 260 U. S. 393, 413; 43 Sup. Ct. 158, 159; 67 L.ed. 322, 325, (1922).

This quotation is also cited by C. I. Hendrickson, "Rural Zoning: Controlling Land Utilization under the Police Power," *Journal of Farm Economics,* Vol. 18, No. 3, Aug. 1936, pp. 485–486.

affecting practices is given in bovine tuberculosis control. Constitutional obstacles to the extension of this principle apparently do not exist so long as funds from taxation are spent for "public purposes."

Public Management of Resources

If the practices required are very detailed and stringent, or if a large compensation (large relative to other costs of the utilization plan) is necessary for a considerable period, public management is administratively more effective and cheaper in accomplishing objectives of conservation policy than regulation of private enterprises. These two conditions are usually interrelated. With helium, for example, regulation would have to be so detailed and stringent for military reasons that private enterprise could operate only under large compensation. Resources administered by the United States Forest, Grazing, Park, Fish and Wildlife services, the defense establishment, and the Atomic Energy Commission are likewise of a kind that private enterprise could operate only under detailed and stringent regulation or large compensation or both.

Two opposing tendencies make it rather difficult to predict the relative future significance of regulation and outright public management of resources.

On the one hand, economic conditions may change in such a way that certain socially desirable practices may also become privately advantageous or at least less disadvantageous. This may expand the economic limits of regulation. In private forestry, for example, practices now confined to public forests may become economically feasible in private enterprises after the last virgin stands have been liquidated; replacement of donkey and railroad logging by caterpillar and truck has greatly facilitated socially desirable practices. Changes with similar results have occurred in agriculture and mining.

On the other hand, citizens and their governments are becoming more and more concerned about the public values in resource utilization. This awareness of public values will make greater participation of the public in resource management unavoidable —provided that the economic limits of regulation are not expanded.

Direct Tools and the Safe Minimum Standard

It was argued in the preceding chapter that, with critical-zone resources, conservation policy should focus on a safe minimum standard as its first objective, rather than on the social optimum. Such limitation in the objectives of conservation policy appears particularly advisable if the tools are zoning and regulation.

No great harm can come from striving for any higher than the safe minimum standard of conservation if the tool is education. Zoning and regulation, on the other hand, impose an economic burden on private enterprise and from a certain degree onward make the whole system of private enterprise unworkable. The constitutional and statutory safeguards on the employment of zoning and regulation, which were noted above, are largely imposed for this reason. The social "concept" (chapter 16) of such provisions is that zoning and regulation should be employed only under conditions in which the public purpose is fairly clear and compelling. The latter is true if the safe minimum standard is at stake. The public purpose becomes less and less clear and compelling the more ambitious the objective that is adopted for conservation policy.

The same characteristics (direct interference with private enterprise) which make zoning and regulation socially and legally risky if used for more ambitious objectives than a safe minimum standard, make them especially effective for realizing such a standard. As already noted, the results of education, subsidies, and of indirect tools of conservation policy are uncertain in degree, geographic extent, and timing. These tools do not guarantee a safe minimum standard for all resource users in a given area within a certain period of time. Through zoning and regulation, on the other hand, a safe minimum standard can be established in all private enterprises in a given area within any desired period of time. Still, only a few enterprises may be affected adversely and these only to a small extent. Such minimum interference with private enterprise is possible because, as we know, the safe minimum standard is frequently a rather modest state of conservation relative to the private optimum.

CHAPTER

20

INTERNATIONAL TOOLS OF CONSERVATION POLICY

An International Conservation Policy?

In considering the international problems of conservation policy, the first questions that arise are: With reference to what social group may objectives and criteria of conservation policy be defined? What government may be regarded as the agent that employs tools of international conservation policy?

At a certain degree of enlightenment of national groups, and at a certain state of international organization, one may be justified in considering an international conservation policy in the sense that objectives and criteria are defined in terms of a world community, and that the tools of conservation policy are entrusted to international agencies.

Such an international conservation policy is a rather distant possibility. At present, enlightenment of nations with respect to subordinating national to international objectives cannot be said to have advanced very far; international organization and law are at best in a formative state. For the present, therefore, it is more realistic to consider tools of national conservation policy in the field of international relations; for short, they will be called "international tools."

Although one cannot yet speak of an international conservation policy in the first sense, there are many historical cases in which the conservation policies of two or more nations merged with respect to objectives and tools. This is especially true for the conservation of flow resources. Some lessons can be learned from such attempts at international coöperation in conservation policy and from the United States participation in them. This approach to studying the international tools of conservation policy, involving emphasis on experience in international coöperation, will be used in the major part of this chapter.

Another approach may proceed from the conservation objectives of individual nations and may study the most appropriate tools of realizing these objectives in the field of international relations. In this approach, potential conflicts of national interests, rather than coöperation, will receive emphasis. Such conflicts exist especially in the conservation of stock resources. This approach will be used for the United States in the last part of this chapter.

Since it seems likely that the international tools of conservation policy will become even more important in the future than in the past, a fairly detailed discussion of these problems seems justified.

Four Tools of International Coöperation

Previous attempts at international coöperation in conservation policy have employed four major tools. First, officially sponsored international conferences have been employed to exchange information and discuss common problems. Second, treaties (conventions) between two or more countries have been employed to conserve those resources in which there existed a definite common interest. Third, permanent international agencies created by member governments have been employed to gather information and to supervise treaties and agreements. Fourth, agreements between organized national groups of producers (international commodity agreements) have resulted, often unintentionally, in conservation or depletion. National groups of producers were usually aided or guided by their governments, and governmental agreements frequently grew out of private ones. Such agreements are, therefore, legitimately treated as part of public policy.

These four major tools are closely related. International conferences were usually the first step toward treaties. Treaties and agreements, in the field considered here, often set up an international agency for facilitating or supervising their operation.

The earliest, most numerous, and most successful attempts to use international tools in conservation policy were concerned with individual fugitive resources in limited areas. These attempts will be discussed first. More ambitious but as yet less successful attempts have had wider scope with respect both to the number of resources and to geographic extent. After discussing the experience gained from these attempts, we will deal with conservation potentialities of international commodity agreements.

International Coöperation in Fugitive Resources

The fugitive resources most often of interest from the international point of view are fish, other marine life, and birds —although upland game, oil, gas, and ground water occasionally raise international issues of a similar nature. These resources must be captured before use. The ones that are flow resources have a critical zone, and a safe minimum standard of conservation should, therefore, be the base level for international action.

There are two regulatory approaches to realize such an objective. First, there is the possibility of regulating capture by international conventions providing for sanctuaries, closed seasons, limits with respect to number, size, sex, age, and by prohibiting certain implements of capture. Second, there is the possibility of safeguarding national control of fishing or hunting grounds in such a way that domestic tools of conservation policy can be applied; this means the establishment of geographically limited national monopolies.

The early importance of the doctrine of freedom of the seas (Hugo Grotius' *Mare Liberum*, 1609), British interest in this doctrine as well as British naval power to enforce it, and the tapping of new resources with new techniques as older resources were depleted postponed international regulation on the high seas until the end of the nineteenth century. Up to that time many treaties aimed at safeguarding national monopolies in marine resources through interpretation of territorial waters.[1] The most extreme and most recent attempts in this direction are the doctrine of the "continental shelf" advanced by some groups in the United States and the 200-marine-mile limit provided by the constitution of El Salvador of 1950.

Conferences and Conventions on Marine Resources

When the age of international regulations began, fisheries conventions (the North Sea Convention, 1882; the Spitsbergen Convention, 1920; the Baltic Sea Convention, 1929; the Geneva Whaling Convention, 1931; and the London Whaling agreements, 1937, 1938, and 1939) were closely related to and facilitated by the discussions at international conferences. The first was the International Fisheries Exhibition in London in 1883, held as a sequel to the newly concluded North Sea Convention. The first international fisheries congress met in 1896, followed by others in 1898, 1899, 1901, and 1902. At the last-mentioned conference, in Copenhagen, the oldest and most successful international agency connected with conservation was established. This was the International Council for the Study of the Sea.[2] At the turn of the century a new series of conferences was begun on a still broader international basis. The first international Congress of Agriculture and Fisheries, at Paris in 1900, advised establishing

---

[1] Stefan A. Riesenfeld, *Protection of Coastal Fisheries under International Law* (Washington: Carnegie Endowment for International Peace, 1942), 296 pp. (Monograph Series of the Carnegie Endowment for International Peace, Division of International Law, No. 5).

[2] Established by the governments of Denmark, Finland, Great Britain, Germany, The Netherlands, Russia, and Sweden, with headquarters in Copenhagen. The Council published many volumes of proceedings, reports, and periodical bulletins connected with hydrography, meteorology, plankton, fishing technology, and regulations. Similar organizations of smaller importance are the International Commission for the Study of the Mediterranean, the International Commission for the Study of the Atlantic, both established in Monaco in 1910, and the North American Council on Fisheries Investigations established in Ottawa in 1920. See Philip C. Jessup, "Exploitation des Richesses de la Mer," *Recueil des Cours* (Académie de Droit International de la Haye), Vol. 29, 1929, pp. 405–514.

a permanent international commission to regulate fisheries. Nothing, however, came of this proposal at the following congresses. Only for whaling did the Geneva Convention of 1931 establish such an agency, the International Bureau of Whaling Statistics at Oslo.[3]

The United States and Argentina were parties to the whaling conventions. This was not the first step in the New World toward international coöperation in conserving marine resources. After a twenty-year diplomatic history (starting with the United States–British treaty of 1891) the United States, together with Great Britain, Russia, and Japan, in 1911 signed the International Convention for the Protection of Fur Seals.[4] Since fur sealing on shore is a United States monopoly (pelagic sealing was prohibited as part of the convention), this country was given the full right to enact regulations for resource conservation. Thus, the United States government, like the French government in the Paris Convention mentioned later, may be regarded as the convention's international agency. The Fur Seal Convention operated with excellent success until its abrogation by Japan in 1940, which became effective one year later—a few weeks before Pearl Harbor. In 1942 an agreement between the United States and Canada was substituted for the convention, and negotiations for a new international Fur Seal Convention are now in progress.

---

[3] For more details about this and about the international whaling conferences in London (1937, 1938, 1939) see Larry L. Leonard, "Recent Negotiations toward the International Regulation of Whaling," *American Journal of International Law*, Vol. 35, No. 1, January 1941, pp. 90 ff.

Whereas the international whaling agreements were largely concerned with the conservation of the resource itself, the "whaling cartel" concluded between private British and Norwegian interests and operating at various periods between the Geneva and London agreements aimed mainly at orderly marketing in the interest of high and stable prices. The results, however, were probably more helpful for conservation than the Geneva convention, which was slow in being ratified and put into effect by its signatories.

[4] An account of the history of conservation of marine resources in the New World is given in Jozo Tomasevich, *International Agreements on Conservation of Marine Resources, with special reference to the North Pacific.* (Stanford University, Calif.: Food Research Institute, 1943), 297 pp. (Food Research Institute Commodity Policy Studies No. 1).

Conservation of Pacific Halibut and Salmon Fisheries

Of special interest, for our present purpose, are the joint efforts of the United States and Canada to conserve Pacific halibut resources. After an unsuccessful attempt during the American-Canadian Fisheries Conference of 1918, a treaty was signed in 1923 and promptly ratified. It provided for closed seasons and for a permanent International Fisheries Commission, at first charged with establishing the facts on which regulation would be based. The treaty was revised in 1930 and 1937. In 1930 the commission received full regulatory powers, subject to the approval of the President of the United States and the Governor-General of Canada.

Two features of halibut conservation deserve special attention here. The first is the establishment of an advisory conference board representing all interests of the industry. The second is the operation of the Halibut Production Control Board, essentially a private cartel of the American halibut industry, operating with the unofficial approval of the commission and with the coöperation of Canadian interests. The conference board has proved useful for obtaining the voluntary coöperation of the industry in executing the commission's regulations.

The control board has enabled a kind of division of labor between the commission and private industry. The commission regulates the industry strictly in the interest of conservation (sets annual quotas of catch, establishes closed seasons, restricts fishing in nursery grounds, regulates fishing gear, collects statistics, and the like). The board attempts orderly marketing throughout the year in the interest of prices and employment.

It may be noted that the commission does not supervise the operation of the private cartel (although such supervision was repeatedly proposed), provided these operations do not conflict with the commission's regulations. Undoubtedly, however, the quota system administered by the commission and the official statistical information collected by it have greatly facilitated the operation of the cartel.

The problems of the Pacific sockeye-salmon industry are more complex than those in halibut conservation. The problems are biological, technical, and political. The diplomatic history of the American-Canadian Sockeye Salmon Convention of 1930, starting with the American-British agreement of 1892, is even longer than that of the Fur Seal Convention. The convention established an International Pacific Salmon Fisheries Commission for scientific investigations and regulations. It also provided for an advisory committee representing private industry. In these and in other ways the convention followed the good experience gained in halibut conservation. The United States did not ratify the convention until 1937. Investigations by the commission began in 1938. Regulation started much later, after investigations had covered two cycles of sockeye-salmon runs—that is, a period of eight years.

The conventions with Canada regarding the Pacific halibut and salmon are mentioned as illustrations because United States experience in these cases extends over a number of years, and because regulation has actually been accomplished. Several other fisheries conventions have been concluded by the United States with her neighboring countries which provide for a permanent international commission for study and consultation, but not yet for regulation. Recent examples are the conventions between the United States and Mexico of January 25, 1949, and between the United States and Costa Rica of May 31, 1949, for the scientific investigations of tuna.

Conferences and Conventions on Migratory Birds

The same close connection between international congresses, conventions, and permanent agencies exists for fugitive resources other than fish. The first International Ornithological Congress, held at Vienna in 1873, appointed an unofficial committee of experts to study the international protection of migratory birds. Similar proposals were made at the second congress (Budapest, 1891) and at the third (Paris, 1901). Finally (at Paris in 1902), a convention was signed to protect "birds useful in agriculture." The detailed provisions of this convention became effective in 1905 after eleven signers had ratified it.[5] The convention charged the French government with certain permanent functions; for example, exchange of information about the regulations enacted by signatory countries. Although this convention outlasted the war of 1914–1918, its execution by some of the countries was extremely lax.[6] Later international conferences [7] proposed strengthening the Paris Convention. Neither these proposals nor the attempt to make the International Institute of Agriculture in Rome the permanent agency of a new convention met with success.

In the New World, international efforts at bird conservation came about as a legalistic device for meeting the technicalities of the Constitution of the United States, since wildlife is under the jurisdiction of the states. After previous federal attempts to regulate the killing of migratory birds had failed, the treaty solution was proposed in Congress in 1913.[8] A treaty with Canada was signed in 1916. Regulation and enforcement were left to national legislation enacted by the United States in 1918. On the basis of this act, the Migratory Bird Commission was created in 1929 and charged with the selection of sanctuaries. The commission is interesting from the standpoint of policy coördination because it is composed of representatives of several executive departments (secretaries of agriculture, commerce, and interior) and two representatives each of the two legislative bodies. In 1936, a similar treaty was signed with Mexico.

---

[5] Signers were Austria-Hungary, Belgium, Germany, Greece, France, Luxembourg, Monaco, Portugal, Sweden, Switzerland, and Spain. A detailed account of the provisions and of the diplomatic prelude to this convention, especially of the role of Italy (Austrian-Italian declaration of 1891), is contained in Sherman S. Hayden, *The International Protection of Wild Life; an Examination of Treaties and Other Agreements for the Preservation of Birds and Mammals* (New York: Columbia University Press, 1942), 246 pp. (Columbia University Studies in History, Economics and Public Law No. 491).

[6] Article 4 provided that the powers may proceed "cautiously" and "by degrees."

[7] International Congress for the Study and Protection of Birds (Geneva, 1927), International Ornithological Congress (Amsterdam, 1930), Second International Congress for the Protection of Nature (Paris, 1931), Northern Nations Conference (Brussels, 1935), Counseil International de la Chasse (Vienna, 1937).

[8] J. C. Phillips, *Migratory Bird Protection in North America* (New York: American Committee for International Wildlife Protection, 1934), 38 pp. (American Committee for International Wildlife Protection, Special Publications, Vol. 1, No. 4).

Conferences and Conventions on Wildlife in General

The last-mentioned treaty (Article 5) forbids the import and export of upland game animals. The first step toward international conservation of such animals in the New World was shortly followed (in 1940) by the signing of the Pan-American Convention [9] for protecting native flora, fauna, scenery, and objects of aesthetic, historic, or scientific value. This convention was concluded according to a proposal of the eighth Pan-American Conference, held in Lima in 1938. The Pan-American Union is designated as the permanent agency for distributing information. Regulation and enforcement are left to independent action of member governments. Other conservation activities are taken up later in this chapter.

The Pan-American Convention is broader than the London Convention of 1933, which aimed to protect the native flora and fauna of Africa and was signed by South Africa, Great Britain, Belgium, Egypt, France, Italy, Spain, Portugal, and the Anglo-Egyptian Sudan.[10] The London Convention designates the British government as permanent agency for mutual information on actions taken and for other diplomatic and technical functions. It follows a recommendation of the International Congress for the Protection of Nature, held at Paris in 1931.

The preceding congress at Bern (1913) also deserves mention because it proposed to make conservation of fugitive resources world-wide, and suggested a permanent International Commission for the Protection of Wildlife.[11] A draft convention establish-

---

[9] Signers are Argentina, Bolivia, Brazil, Chile, Colombia, Costa Rica, Cuba, Dominican Republic, Ecuador, El Salvador, Guatemala, Haiti, Mexico, Nicaragua, Peru, United States, Uruguay, and Venezuela.

[10] The texts of the American and African conventions are reprinted in Hayden, *op. cit.*, Appendices I and II. The same book gives an account of the long history of the London Convention and of its stillborn predecessor, the African Convention of 1900.

[11] This suggestion goes back to the International Zoölogical Congress at Graz in 1910, which set up a committee of experts to study it. This committee, through the government of Sweden, sounded the principal powers on the subject of an agreement. Sixteen nations replied favorably. Only Japan and Roumania were against the plan. The Bern Congress was called to discuss the next step.

ing such a commission was adopted.[12] The commission was to be charged with collecting, classifying, and publishing data relevant to the international protection of nature and with educating the public and governments in problems of conservation. The outbreak of the war in 1914 interrupted this modest beginning. However, international conferences on the protection of nature continued, and recently have led to another permanent agency, the International Union for the Protection of Nature, formed at Fontainebleau in 1948.

Needs for International Conservation of Fugitive Resources

In spite of all these attempts, existing international regulations of fugitive resources leave much to be desired. The United States is especially interested in international regulation of the fisheries of the Great Lakes, the tuna and shark of the Pacific and Atlantic, and the cod and haddock of Newfoundland. These coastal fisheries form an exception to the rule that wars tend towards conservation of marine resources. They were seriously depleted during the Second World War because of geographical location and of increased war demand for liver oils (largely for vitamin-A production) and fish proteins. A similar experience would accompany another national emergency. On the other hand, wars threatening the safety of the high seas tend towards conservation of fur seal and whaling resources.

Existing migratory-bird treaties to which the United States is a partner should be expanded to include all Central and South American countries; and the Pan-American Convention, now merely a lofty aim, should be made more effective. Many species that breed in the north during the summer go south during the winter, some of them all the way to Patagonia.[13]

In the Old World, a similar complementarity exists between

[12] Conférence Internationale pour la Protection de la Nature, *Recueil des Procès-verbaux* (Bern: K. J. Wyss, 1913), 247 pp.

[13] Alexander Wetmore, *Our Migrant Shorebirds in Southern South America* (Washington: U. S. Department of Agriculture, 1927), 24 pp. (U. S. Department of Agriculture Technical Bulletin 26).

Europe and Africa. The Paris Convention, never effective and always narrow in scope and geographical extent, has become obsolete through boundary changes and new technological developments (the advent of automobiles, motorboats, and airplanes). Previously remote areas that functioned as natural sanctuaries are now easily accessible.

International Coöperation of Broader Scope

The first attempt to achieve international coöperation in the conservation of all natural resources of the world proved unsuccessful. This was the attempt connected with the name of President Theodore Roosevelt at the first peak of the conservation movement in the United States.

The North American Conservation Conference at Washington, February 18–23, 1909,[14] carried Roosevelt's spectacular domestic campaign of conservation into the international field.[15] It resulted in a highly interesting Declaration of Principles recommending various domestic conservation policies for land, water, forests, minerals, and wildlife.[16] Some of these recommendations have since been adopted, and most of them are relevant today. Two of them, affecting international tools of conservation policy, fall within the scope of the present survey. One relates to the appointment of conservation commissioners in each country, who, at stated intervals, should communicate to each other all informa-

[14] For this conference three commissioners each were appointed by the President of the United States, the President of Mexico, and the Governor-General of Canada; and one commissioner by the Governor of Newfoundland. The United States commissioners were men of the stature of Gifford Pinchot, Robert Bacon, and James Randolph Garfield.

[15] Domestically this campaign had culminated the year before in the White House Conference of Governors and in the appointment of a National Conservation Commission. See:

Conference of Governors on Conservation of Natural Resources, *Proceedings of a Conference of Governors in the White House, Washington, D. C., May 13–15, 1908* (Washington: Govt. Print. Off., 1909), 451 pp. (U. S. 60th Cong., 2d sess., H. Doc. 1425).

National Conservation Commission, *Report of the National Conservation Commission with Accompanying Papers* (Washington: Govt. Print. Off., 1909), 3 vols. (U. S. 60th Cong., 2d sess., S. Doc. 676, serial nos. 5397–99).

[16] Reprinted in Loomis Havemeyer and Gar A. Rousch, *Conservation of our Natural Resources* (New York: The Macmillan Company, 1937), pp. 535 ff.

tion pertaining to conservation. The second relates to a world conference "on the subject of world resources and their inventory, conservation and wise utilization."

The First International Conservation Conference

Even before the North American Conservation Conference, Theodore Roosevelt informally (aide-memoire of January 6, 1909) had sounded out the major powers of the world on whether they would care to send delegates to an international conservation conference. In the midst of the North American Conference, the secretary of state instructed American diplomatic representatives to ask the major powers to meet at The Hague to discuss conservation for the entire globe. This instruction of Secretary Robert Bacon is dated February 19, 1909; it did not constitute a formal invitation.[17]

This step was taken two weeks before President Taft's inauguration. Congress had for some time adopted a negative attitude toward the problem. Not only was the conservation commission appointed in 1908 left without funds, but the enmity of Congress went so far as to prevent, through a legislative rider, all executive departments from performing work for any commission, council, board, or similar body appointed by the President. Under such unfavorable political conditions, nothing came of a World Conservation Conference, although twenty-nine powers, among them Great Britain, France, and Germany, sent favorable replies. The Department of State, in response to an inquiry of January 13, 1911, from the Netherlands government regarding the issuance of formal invitations, replied that the United States government, "observing the general lack of any interest commensurate with the difficulties of the questions to be considered by the proposed conference," had concluded that it would take no further steps toward holding the meeting. A circular instruction to this effect was also sent to the diplomatic officers of the United States.[18] With

[17] U. S. Department of State, *Papers Relating to the Foreign Relations of the United States, 1909* (Washington: Govt. Print. Off., 1914), p. 2. (U. S. 61st Cong., 2d sess., H. Doc. 101).

[18] For this information the writer wishes to thank Mr. E. Wilder Spaulding, Chief, Division of Research and Publications, Department of State.

respect to the failure to proceed with the preparations, it may be noted that original sponsors of the conference, notably President Roosevelt and Governor Pinchot, former head of the U. S. Forest Service, no longer had any official connection with the project.

Political antagonism was not the only reason why Roosevelt's efforts were frustrated. As noted in chapter 1, an idea with a strong emotional appeal, and with an economic and social rationale more difficult to grasp than might appear at first sight, may be used for and compromised by objectives which are not necessarily in agreement with it. For conservation such objectives have been at various times the breaking of monopoly power, stabilization of prices, increase of public control over resources, increases in employment, and changes in the income distribution. In Theodore Roosevelt's era, conservation became identified especially with the first of these objectives. Those who disagree with this historical identification have said, "the trust-buster and the conservationist are strange bedfellows." [19] This is not necessarily true. Conservation policy need not conflict with an antimonopoly policy (chapter 14). On the other hand, an uninformed, partisan anticartel attitude combined with vague, emotional notions about conservation confuses issues, discredits bona fide conservation, and makes it difficult to employ tools of conservation policy that are potentially effective and socially not necessarily harmful. The parallelisms between the earlier conservation movement and the present are striking; they will warrant attention below in connection with international commodity agreements.

Pan-American Conservation of Flow Resources

The domestic setbacks of the conservation movement in the United States slowed, but did not stop, its spread over the New World. In accordance with Roosevelt's ideas and in contrast to the situation in Europe, coöperation among American countries did not remain confined to fugitive resources but is now concerned with most flow resources. The inclusion of scenic and cultural resources in the agenda of the Pan-American Convention of 1938

[19] Erich W. Zimmermann, *World Resources and Industries* (New York: Harper and Brothers, Publishers, 1933), 785 pp.

was mentioned above. An even broader attitude was taken by the Pan-American Scientific congresses (not to be confused with the Pan-American Conference mentioned above in connection with the Pan-American Convention). Especially noteworthy in this regard were the fifth and the eighth congresses, held in Washington, D. C., in 1915 and 1940.[20]

With respect to policy recommendations, the fifth congress confined itself to coöperation in plant and livestock protection and to a pan-American study of forest resources.

The eighth congress went further. It recommended that the governments of the American republics appoint an Inter-American Conservation Commission to be charged with the duty "of preparing an inventory of world natural resources and of formulating a general policy and specific program of action to promote the mutual conservation and prudent utilization of natural resources for the welfare of all nations in the interest of permanent peace." [21] Although these recommendations were not acted upon—a common fate of such resolutions by international conferences—the following decade saw some improvement in international coöperation among the American nations in the conservation field.

The Pan-American Union, through its Conservation Section,

---

[20] In these congresses the Latin American countries had taken the leadership. The first three were held at Buenos Aires in 1898, at Montevideo in 1901, and at Rio de Janeiro in 1905. The fourth, at Santiago in 1908, became the first Pan-American Congress through a strong delegation from the United States. Since then the American Scientific Congress, especially the fifth, held in Washington, D. C., in 1915, and the eighth, held there in 1940, have given prominent attention to conservation. Previously, only individual papers had dealt with the subject. See *Report of the Delegates of the United States to the Pan-American Scientific Congress held at Santiago, Chile, December 25, 1908 to January 5, 1909.* (Washington: Govt. Print. Off., 1909), 65 pp.

The proceedings of the fifth congress contain a large volume on conservation of natural resources, agriculture, irrigation and forestry.—Pan-American Scientific Congress, 1st, Santiago, Chile, 1908, *Proceedings* (Washington: Govt. Print. Off., 1917), Vol. 3.

The proceedings of the eighth congress contain a volume on agriculture and conservation.—American Scientific Congress, *Proceedings of the Eighth American Scientific Congress, held in Washington, May 10–18, 1940* (Washington: U. S. Department of State, 1942), Vol. 5.

[21] This recommendation follows closely a suggestion by Gifford Pinchot in his paper presented before the Congress. See American Scientific Congress, *ibid.*, p. 23–28 and Vol. 1, pp. 244 ff.

sponsored and published a number of studies on various South American countries.[22]

The Pan-American Union also took the initiative in convening the Inter-American Conference on Conservation of Renewable Natural Resources in 1948.[23] This conference passed a number of resolutions relating to inter-American coöperation in the conservation field. The main contribution of the conference, however, was that for the first time some economic, social, and political issues on resource conservation were frankly discussed. Two sections (Section II, Renewable Resources and International Relations; and Section III, Land Use and the Social Sciences) considered these issues.

In soil conservation, several South American countries have established agencies patterned after the United States Soil Conservation Service. The latter has helped with advice, training, and exchange of personnel.

More than advice is given by the Institute of Inter-American Affairs, established within the government of the United States for "collaboration with other governments and governmental agencies of the American republics in planning, initiating, assisting, financing, administering, and executing technical programs and projects especially in the fields of public health, sanitation, agriculture, and education." [24] Agricultural experiment stations are maintained through coöperation between the United States Department of Agriculture and South American governments. An Inter-American Institute of Agricultural Science has been established at Turrialba, Costa Rica.

---

[22] William Vogt, *The Population of Venezuela and Its Natural Resources* (Washington: Pan-American Union, Division of Agricultural Cooperation, 1946), 52 pp. Processed.

William Vogt, *The Population of El Salvador and Its Natural Resources* (Washington: Pan-American Union, Division of Agricultural Cooperation, 1946), 30 pp. Processed.

William Vogt, *The Population of Costa Rica and Its Natural Resources* (Washington: Pan-American Union, Division of Agricultural Cooperation, 1946), 25 pp. Processed.

[23] U. S. Department of State, *Proceedings of the Inter-American Conference on Conservation of Renewable Natural Resources, Denver, 1948* (Washington: 1949), 782 pp. (U. S. Department of State Publication 3382. International Organization and Conference Series II, American Republic 4).

[24] 69 *U. S. Stat. at L.* (1947), p. 780.

Conservation of Flow Resources through World Organizations

For the world as a whole, coöperation in the conservation field is much less than among the American republics. Among the relevant international agencies, the International Institute of Agriculture in Rome, founded in 1902 by an international convention, had a singular opportunity to take world leadership in research and policy with respect to land, water, forest, and wildlife conservation. It has facilitated a few international conventions for plant protection (for example, the International Convention on Locust Control of 1920; International Convention on Plant Protection of 1929). Some of its data on international movement of agricultural products and fertilizers are valuable. But its voluminous studies on agricultural production, trade, and law scarcely mention the technological, economic, social, and political aspects of conservation. The International Forestry Center in Berlin, a branch of the institute established in 1937, had just started to function when the Second World War broke out.

The functions of the International Institute in Rome have been taken over by the Food and Agriculture Organization (F.A.O.) of the United Nations. International coöperation in the conservation field is mentioned among the objectives of the F.A.O. in its constitution. It has published an international survey of soil conservation.[25] More recently, through conferences and missions to underdeveloped countries, it has taken a strong interest in the important field of water-resources development. The F.A.O. participated in the preparation of the United Nations Scientific Conference on Conservation and Utilization of Resources which convened at Lake Success in 1949.[26] This conference, unlike the Inter-American Conference on Conservation of 1948, failed to include any study of the essential economic, social, and political problems of conservation. The F.A.O. has failed to make a con-

[25] United Nations, Food and Agriculture Organization, *Soil Conservation; an International Study* (Washington, 1948), 189 pp. (F.A.O. Agricultural Studies No. 4).

[26] The proceedings of this conference (8 volumes) are in process of being published.

tribution in a field for which it is particularly well equipped—namely, in undertaking a thorough and impartial analysis of the conservation aspects (among others) of international commodity agreements.[27]

Conservation of Stock Resources through World Organizations

For flow resources the F.A.O. provides an organization that is taking some steps to improve international coöperation in the conservation field. No such organization is provided within the United Nations for stock resources. The technological, economic, and social problems connected with international conservation of minerals are just as important and just as complex as similar problems related to agriculture, forestry, and fisheries. As discussed in more detail in the subsequent sections of this chapter, conflicts of interests between nations and organized private groups are much stronger with stock than with flow resources. A Mineral Resources Organization of the United Nations is needed fully as much as a Food and Agriculture Organization. It would not compete with the International Trade Organization but would be complementary to it in fact-finding, research, and advisory functions (see footnote 27).

An International Resources Organization has been proposed by the Division for the Social and International Relations of Science, in the British Association for the Advancement of Science.[28] The report of the Mineral Resources Committee ap-

[27] It may be objected that international commodity agreements are the responsibility of the International Trade Organization (I.T.O.). This may be true; however, a separation of operation and investigation in this field between the I.T.O. and the F.A.O. may not be at all harmful. The operation of commodity agreements is exposed to strong pressure by governmental and private groups. The personnel best suited for investigation may be less well suited for operation. A division of labor between operating and fact-finding functions has proved beneficial in domestic government (chap. 21).

[28] British Association for the Advancement of Science, "Science and World Order: Transactions of a Conference of the Division for the Social and International Relations of Science," *Advancement of Science*, Vol. 2, No. 5, January 1942, pp. 3–116.

British Association for the Advancement of Science, "Mineral Resources and the Atlantic Charter: Transactions of a Conference held by the Division for the Social and International Relations of Science, July 24–25, 1942," *Advancement of Science*, Vol. 2, No. 7, October 1942, pp. 187–253.

pointed by the division to consider this suggestion confines its recommendations, however, to national minerals and metals resources boards.[29] Such boards exist in some form in most countries. Although useful, they are not sufficient for international coöperation. What is needed is an international organization which may act as a clearinghouse for information collected by national resources boards; which may be available as an investigating and advisory body to governments and international organizations (the I.T.O., the International Bank, the International Court, the Security Council); which may initiate and stimulate international action; and through which nations with conflicting interests may learn to coöperate on a fact-finding rather than a judiciary or political level.

This last point needs emphasis. It appears doubtful that military conflicts between nations are caused mainly by economic factors, especially by division into "haves" and "have nots" through the geographical distribution of natural resources. Historically, the relation between political and military conflict and economics is of great importance; but in this relation the economic factor plays a passive and conditioning role more often than is generally supposed.[30] Unquestionably, on the other hand, the world distribution of stock resources in an industrial age decisively influences the political ambitions and fears of nations. There is scarcely a field in which international institutions and habits of coöperation could pay larger dividends for the maintenance of world peace than the field of stock resources.

Potentialities of International Commodity Agreements

International agreements concerning the production and marketing of the primary products of extractive industries—for

[29] British Association for the Advancement of Science, "Mineral Resources and the Atlantic Charter: Report of the Mineral Resources Committee," *Advancement of Science,* Vol. 2, No. 8, August 1943, pp. 339–345.

[30] S. V. Ciriacy-Wantrup, "Agrarkrisen und Stockungsspannen zur Frage der langen 'Welle' in der wirtschaftlichen Entwicklung," [Germany] Reichs- und Pr. Ministerium für Ernährung und Landwirtschaft, *Berichte über Landwirtschaft,* N. F. 122. Sonderheft (Berlin: Paul Parey, 1936), 445 pp. Also: *Agricultural Depressions and Periods of Industrial Stagnation,* translated by C. O. Hardy (Washington: Brookings Institution, 1938), 296 pp. Processed.

short, commodity agreements—may be concluded between individual firms, between groups of firms, or between governments. In contrast to patent agreements,[31] commodity agreements have more and more been concluded between governments or, at least, with the encouragement, advice, or supervision of governments. This trend can be expected to continue.

Most of the argument for and against the social desirability of commodity agreements concerns their effectiveness in reducing price instability. If agreements were effective in this respect, they would indeed have significance for conservation. The reasons were explained previously: First, greater price stability may reduce uncertainty allowance and, therefore, result in conservation. Second, greater price stability may reduce interest rates and capital rationing because uncertainty allowance of savers and lenders is decreased; lower interest rates and less imperfect markets for loans favor conservation. Third, greater price stability may avoid depletion caused by low incomes during economic depressions under imperfect markets for assets; this depletion is not offset by conservation during prosperity because then markets for assets are less imperfect and variations of incomes have less and less influence upon time preference as incomes increase.

Reasons were given in chapter 15 why commodity agreements are not likely to reduce price instability sufficiently in degree and duration to be of interest from the standpoint of conservation. Experience with commodity agreements supports this view. We concluded that the conservation argument for commodity agreements must be built on a stronger foundation. A foundation was sketched: commodity agreements have potentialities as international tools of conservation policy if they could be used (1) to control expansion of production under conditions where such expansion can be only temporary and where difficulties of con-

[31] Patent agreements are related to production because they may be used for developing or not developing resources. In actuality, however, they are more important for manufacturing. Price agreements are generally a part of an agreement that allocates markets. A sharp economic differentiation between commodity agreements and cartels, which currently is in vogue, appears questionable logically as well as practically.

traction lead to wasteful depletion; (2) to help in the orderly contraction of production under conditions where the death struggle of temporarily submarginal capacity leads to wasteful depletion; (3) to facilitate international adoption of conservation practices.

Such potential usefulness of commodity agreements for conservation policy was considered in relation to economic instability, defined in terms of shifts of demand. However, problems of overexpansion and slow contraction occur not only in the course of these shifts: important technological changes (for example, the advent of railroad and steamship, the mechanization of agriculture through tractor, harvester, and truck), decisive changes of institutional and political conditions (for example, the Homestead Law in the United States, opening and closing of access to resources by war and conquest), and discovery of new stocks of resources may create similar problems. A few broad illustrations from agriculture, forestry, and mining may be helpful at this point; more details will be given later when historical experience with commodity agreements is reviewed.

In agriculture, mechanization may make profitable—at least privately and temporarily—a sudden and vast expansion of cultivated acreage in areas which should remain in grass if their permanent productivity is to be assured and if damage to other areas through distant effects of accelerated wind and water erosion is to be avoided. At the same time, such overexpansion of cultivated acreage may cause decreases of product prices which may make conservation practices unprofitable in older farming regions—ones that may well be permanently suited for cultivation under proper conservation practices. Under such conditions, control of expansion of cultivated acreage would eliminate a multiple cause of wasteful soil depletion.

In forestry, expansion of cutting in virgin stands because of changes in technology, in accessibility, in social institutions, and in foreign relations may cause such low prices that cutting methods and other silvicultural practices which are necessary for proper restocking become unprofitable. Such practices may remain unprofitable until virgin stands are exhausted. At that time an acute scarcity of timber may arise and persist for a long time until second growth can take the place of virgin stands. Through more orderly cutting of virgin stands and by using the higher returns for conservation of new growth, the violence of ups and downs in timber supplies could be mitigated.

In mining and oil drilling, discovery of resources with especially low development and production costs may temporarily render the exploitation of "old" deposits unprofitable. If such old deposits are abandoned without provision for adequate maintenance, resources may be permanently lost which would not be submarginal after the cream of the new discoveries is skimmed off. A more orderly exploitation of the new discoveries, together with provisions for proper maintenance of older deposits on a stand-by basis, would avoid such wasteful depletion.

Domestically, these problems of overexpansion and slow contraction may be attacked by each country with a number of indirect and direct tools. Internationally, an attack requires agreements with other countries for a coördinated use of these tools.

Experience with International Commodity Agreements

Thus far, we have suggested merely that commodity agreements have potentialities for conservation policy. We may now inquire whether such potentialities have actually been realized and, if not, what the obstacles were and whether such obstacles could be overcome in the future. In considering the historical experience it is well to confine ourselves to those commodity agreements which were in operation between the two world wars. Commodity agreements concluded *during* the Second World War were dominated by political and strategic considerations. Commodity agreements concluded *after* the Second World War have not operated long enough to permit their evaluation. First, we will discuss commodity agreements for flow resources and then for stock resources.

Timber

For flow resources, commodity agreements relevant here are those concerning timber (agreements of 1933 and 1934, Copenhagen Convention of 1935), rubber (agreements of 1934 and amendments), and sugar (agreements of 1931 and 1937). The tea agreement of 1933 was neutral with respect to conservation.[32] The International Wheat Agreement of 1933 broke down in its first year of operation. The various Brazilian coffee valorization schemes which preceded the Inter-American Coffee Agreement of 1940 were not commodity agreements.

The European Timber Exporter's Convention (E.T.E.C.), concluded at Copenhagen in 1935, and its predecessors, the gentleman's agreements of 1933 and 1934, sponsored by the International Timber Committee (C.I.B.), provided for export quotas and voluntary production restrictions of Europe's softwood lumber industry.[33] Safety clauses made it possible to exceed quotas by 10 to 15 per cent, and enforcement was rather lax. There were no provisions for better silvicultural practices. The recovery of European lumber prices in the second half of the 1930's was largely due to demand factors. The E.T.E.C. at least did little to impede this recovery. Export quotas were instituted only after consultation with importing countries. Price policies were directed more at eliminating short-run price fluctuations, through coördinating the marketings of the various exporting countries, than at increasing prices through over-all restrictions.

Although the objectives of the E.T.E.C. were not ambitious and

---

[32] Although the tea plant needs good drainage and is therefore often grown on sloping land, it is usually pruned to a bushy form that affords effective soil protection. Tea requires heavy fertilization. Growing practices include the interplanting of shade trees and green-manure crops. Restrictions of production are achieved through finer plucking. Restrictions were mild, since the main producer and consumer interests belonged to the same country.

[33] As an outcome of studies by the League of Nations undertaken at the suggestion of the Geneva Economic Conference [*The Timber Problem, Its International Aspects*, L. of N. Publications Series II, Economic and Financial, 1932, II.B.6. (Geneva: League of Nations, 1932)], the International Timber Conference at Vienna in 1932 formed the Comité International de Bois (C.I.B.). This organization comprised originally Austria, Latvia, Poland, Roumania, Czechoslovakia, and Yugoslavia. France joined in 1933 and other countries later. The C.I.B. served mainly to collect international statistics on supply and demand of timber. It also sponsored the International Conference of Timber Exporters at Berlin in 1933 and at Vienna in 1934, which led to the gentleman's agreements mentioned in the text. The formal convention in the following year was ratified by Austria, Finland, Poland, Roumania, Sweden, Czechoslovakia, and the U.S.S.R. The E.T.E.C. accounted for 95 per cent of Europe's softwood lumber exports.

its accomplishments with respect to conservation insignificant, the convention may be regarded as a beginning in the right direction. As already implied, a commodity agreement comprising exporting and importing countries would be helpful in solving several important conservation problems in forestry. Sudden liquidation of timber stands, because of financial strain and need for foreign exchange, or because of changes in transportation facilities, in technology, and in political relations, may temporarily cause such low prices that those cutting methods and other silvicultural practices which are necessary for proper restocking become uneconomical. Depletion of immature growing stock may be encouraged by the same factors and also by high prices—which under certain conditions [34] follow low prices caused by rapid liquidation.

A commodity agreement which would provide for orderly liquidation of mature forests, which would prevent depletion of immature stands, and which would spread practices and institutions (in the fields of tenure, taxation, and credit) suitable for sustained-yield management, appears desirable from the standpoint of consumers as well as producers. Although forest products are industrial raw materials and, therefore, subject to violent changes of demand in the course of economic fluctuations, they offer good opportunities for collective control of production and marketing: the standing forest itself provides a large "buffer stock" which, under appropriate systems of tenure, taxation, and credit can be cheaply carried and adjusted.

Rubber

The International Rubber Regulation Agreement (I.R.R.A.) of 1934 was moderately successful in terms of its own objective, to increase net revenues of rubber estates (not necessarily of native holdings).[35] Whether it reduced the vio-

[34] For example, under the conditions prevailing in the United States before the Second World War. The problem of proper timing in liquidating virgin timber in the United States is discussed in S. V. Ciriacy-Wantrup, "Multiple and Optimum Use of Wild Land under Different Economic Conditions," *Journal of Forestry*, Vol. 36, No. 7, July 1938, pp. 665–674.

[35] Little needs to be said here about the history of the I.R.R.A. (which, unlike its predecessor, the Stephenson Plan, 1922–1928, was an international treaty between

lence of price fluctuations is doubtful; in any event, success in this direction was too small to affect resource conservation. The main reason for this failure was that a system of buffer stocks was adopted too late (1938) and on an insufficient scale. The means of obtaining desired objectives were export and production quotas backed by restrictions on replanting and new planting. These means were related to soil conservation in various ways.

First, the distribution of quotas among producers and the restrictions on replanting and new planting favored the estates at the expense of native holdings.[36] Rubber production on estates often results in accelerated erosion, because of clean-row cultivation. Native holdings resemble the natural soil-protecting forest with a much larger number of irregularly planted trees per acre (frequently three times those on estates) and small effort to weed or cultivate.[37] Discrimination against native holdings, therefore, tended toward soil depletion.

Second, no attempt was made to eliminate permanently submarginal capacity. Marketability of quotas was the only, and a rather weak, inducement to curtail production of high-cost producers to a greater extent than that of low-cost producers. Discrimination against native producers tended in the opposite direction: their lower quality and lower price of product are more than offset by lower investment, lower hired-labor costs, and more flexible family-labor costs. Preservation of high-cost capacity accentuates the danger of soil depletion during depressions.

---

signatory governments), its operation, and those results which are not of special interest for conservation. The extensive literature on the subject is fairly well covered by two studies (the first in favor, the second in opposition):

Sir Andrew McFadyean, ed., *The History of Rubber Regulation, 1934–1943* (London: George Allen and Unwin, Ltd., 1944), 239 pp.

K. E. Knorr, *World Rubber and Its Regulation* (Stanford, Calif.: Stanford University Press, 1945), 265 pp.

Neither of these studies inquires into possible relations between soil conservation, rubber culture, and the I.R.R.A.

[36] For data on this point see Knorr, *op. cit.*, pp. 119–124.

[37] In recent years this system of rubber "forestry," which for a long time was looked down upon by European experts as being due to ignorance and laziness of the natives, is spreading among estates. The technological aspects are discussed in Loren G. Polhamus, "Rubber Production as Good Land Use in Tropical America," *Proceedings of the Eighth American Scientific Congress, held in Washington, May 10–18, 1940* (Washington: U. S. Department of State, 1942), Vol. 5, pp. 179–188.

Third, the agreement made little use of the opportunity to diversify farming systems through greater emphasis on food production for the local market. One serious land-use problem in plantation agriculture is the tendency to push native peoples and their food production out of the plains, where mechanization is profitable, into the hills, where mechanization must contend with physical and economic obstacles.[38] In the hills, soil erosion makes rapid progress under the high precipitation (80 to 100 inches) characteristic of rubber regions and under prevailing native practices, especially periodic burning and shifting of cultivation.[39]

Fourth, the agreement offered the possibility of using labor (set free by the restriction) for soil-erosion control and other forms of land amelioration. Labor on estates was largely imported from China and India, and, in the short run, was fixed in terms of numbers. Retention of the labor force was encouraged by year-to-year fluctuations of quotas. In the official history of the I.R.R.A., it is claimed that estates were improved under the agreement.[40] It is impossible to state precisely how significant this conservation effect of the agreement actually was.

The foregoing four points indicate in what way a rubber agreement could be beneficial for resource conservation in the future. The high-cost segment of the estate rubber industry is (or will

---

[38] In some areas this tendency is neutralized by public land policies, for example, in Java by the agrarian decrees of 1870. On the east coast of Sumatra and elsewhere, however, the statement in the text applies.

[39] These practices are not objectionable from the standpoint of soil conservation, if *ladangs* are quickly reforested and if sufficient time is allowed for forest vegetation to restore soil fertility. The latter is a question of the man–land ratio. Under the conditions of rubber areas, it is estimated that serious soil erosion results if the population of shifting cultivators exceeds fifty persons per square kilometer. The impact of Western civilization on native cultures has caused this "critical" ratio to be reached in many areas. After this ratio is reached, shifting cultivation must be replaced by a more stable form of land use.

For illustration from the Asiatic tropics see Karl J. Pelzer, *Pioneer Settlement in the Asiatic Tropics* (New York: American Geographical Society, 1945, Special Publication No. 29), 290 pp. (The Johns Hopkins University Studies on Land Utilization and Agricultural Colonization in Southeastern Asia).

In European land-use history similar problems have been met with. For illustration see S. V. Ciriacy-Wantrup, "Soil Conservation in European Farm Management," *Journal of Farm Economics*, Vol. 20, No. 1, February 1938, pp. 86–101.

[40] Sir Andrew McFadyean, ed., *op. cit.*, p. 150.

soon be) obsolete, because of the quantitative, qualitative, and economic advancement of synthetic rubber, and because of changes in the political, social, and economic institutions of the rubber regions. If the unavoidable adjustments in estate rubber production are to proceed without wasteful depletion of soil resources (and of human and cultural resources), a collective effort will be needed even more than in the 1930's.

It follows from the preceding paragraphs that such an effort will have to take a much broader view of stabilization than the I.R.R.A. did. A buffer stock will probably be less important in the future than in the past, because of great flexibility in the output of already established synthetic capacity—assuming that no fixed ratio between synthetic and natural rubber in the main products will be necessary for technological reasons. On the other hand, consideration of the issues raised above will be vital for success. Such issues were: reversal of discrimination against native holdings in production quotas, elimination of high-cost estate producers, diversification and improvement of remaining estates, and emphasis on food production for the local market.

Sugar

The sugar agreements of 1931 and 1937 established export quotas for sugar-beet and sugar-cane countries. The objective was to enable Cuba and other cane-sugar producers selling in the open market—that is, outside of empire preferences—to make an orderly decrease of production. (Production had expanded greatly during the First World War, when European sugar-beet production had dropped to less than one-third of its former amount.[41]) The effect of the 1931 agreement was largely offset by increases in sugar-beet production for home consumption and of sugar-cane production in exporting countries with a

[41] A basis for appraising the results of these agreements is the statistical material in L. B. Bacon and F. C. Schloemer, *World Trade in Agricultural Products. Its Growth; Its Crisis; and the New Trade Policies* (Rome: International Institute of Agriculture, 1940), pp. 107–178.

See also League of Nations, *International Sugar Conference held in London from April 5 to May 6, 1937. I. Text of the Agreement. II. Proceedings and Documents of the Conference.* L. of N. Publications Series II, Economic and Financial, 1937, II.B.8. (Geneva: League of Nations, 1937), 83 pp.

closed market (mainly members of the British Empire and possessions of the United States). In this respect the 1937 agreement was an improvement, because it set limits to the imperialization of the British and American sugar trade. The agreements did not decrease the violence of price fluctuations; they made no planned use of buffer stocks.

In order to understand the relation between the sugar agreements and soil conservation, one should differentiate between sugar-beet and sugar-cane production. In European agriculture the sugar beet is a valuable crop for increasing soil productivity.[42] Sugar-cane culture, on the other hand, has played an important part in soil depletion, either by expanding into topographically unsuited areas or, in plantation areas, by forcing food production of native peoples to retreat into rough topography. Examples are Cuba, Hawaii, and Puerto Rico.[43] In the native agriculture of these islands, the traditional *milpa* system is still extensively employed.

A sugar agreement which avoids tariff wars in the competition between beet sugar and cane sugar,[44] which prevents expansion of sugar-cane acreage into areas with erosion hazards, which reduces sugar-cane acreage in favor of a more diversified system of farming with emphasis on food production for the local market, and which provides for soil-conservation practices, helps to avoid wasteful soil depletion. The agreement of 1937 tended to operate in these directions. It favored stability in international-trade policies and it aided the United States in limiting the expansion of offshore sugar-cane production (through import quotas) and in

---

[42] S. V. Ciriacy-Wantrup, "Soil Conservation in European Farm Management," *Journal of Farm Economics*, Vol. 20, No. 1, February 1938, pp. 86–101.

[43] Robert G. Bowman, *Soil Erosion in Puerto Rico*, unpublished Ph.D. thesis, University of California, 1941, 187 numb. leaves.

[44] Such a war existed before the Brussels Convention of 1902. This convention abolished export subsidies on beet sugar and limited sugar import duties. Many central European countries employed such export subsidies to increase agricultural productivity (this is an exception to the rule that export subsidies on resources tend to depletion at home). However, subsidizing the export of beet sugar is a wasteful way to obtain soil conservation at home. Moreover, subsidies on beet-sugar exports lead to countervailing duties and to increased assistance to cane sugar by importing countries with a colonial interest in sugar-cane production (British subsidies to the West Indies, assistance to United States possessions).

attacking the problem of diversification in her possessions (through Agricultural Adjustment Administration and Soil Conservation Service legislation). Thus international coöperation proved helpful in finding an orderly solution of the sugar problem.

Stock Resources

In stock resources, agreements have been notably numerous in that group of resources in which abandonment of temporarily submarginal mines without provision for proper maintenance results in wasteful depletion. The economic results of these agreements have frequently been discussed.[45] For this reason, and because, with stock resources, relations between commodity agreements and conservation are rather simple, we can be brief.

Impartial observers agree that during the depression after 1929 many commodity agreements increased producers' net revenues and helped to mitigate the economic pressure toward forced abandonment of mines. Restrictions, however, continued during the following upswing, when they were not needed from the standpoint of conservation.

The record is spotty with respect to planned "rationalization" of an industry through concentrating output in the more efficient enterprises and maintaining the less efficient enterprises in stand-by position. Such a program was carried out successfully through the International Potash Agreement of 1926 and through domestic European potash and coal cartels. Other commodity agreements—for example, the International Tin Agreements of

---

[45] The cartels in the prewar German potash industry are discussed in Robert Liefmann, *Cartels, Concerns and Trusts* (London: Methuen and Co., Ltd., 1932), 379 pp.

The international potash agreements in the postwar period and the nitrate agreements are treated in Benjamin B. Wallace and Lynn R. Edminster, *International Control of Raw Materials* (Washington: The Brookings Institution, 1930), 479 pp.

References to an international oil agreement are found in Herbert Feis, *Petroleum and American Foreign Policy* (Stanford, California: Food Research Institute, 1944), 62 pp. (Food Research Institute Commodity Policy Studies No. 3).

Agreements in bauxite (aluminum) and other nonferrous metals are discussed in:

P. L. Yates, *Commodity Control, a Study of Primary Products* (London: J. Cape, 1943), 248 pp.

K. E. Knorr, *Tin under Control* (Stanford, California: Food Research Institute, 1945), 314 pp. (Food Research Institute Commodity Policy Studies No. 5).

1931 and 1934—encouraged, rather than prevented, the opening up of high-cost deposits. In some quarters [46] such encouragement was defended as socially desirable conservation. Depletion is the more likely result.[47]

Except for limited periods, instability of prices was not reduced, in spite of the fact that buffer-stock devices were employed in a number of agreements. In any event, it is safe to say that possible increases in price stability were not sufficient for resource conservation.

Provision for practices which reduce wasteful depletion were not agreed upon, although they were mentioned in connection with the Anglo-American Oil Agreement of 1944.

Thus, commodity agreements in the field of stock resources, with a few exceptions, have not been successful from the standpoint of this study. Then is the conservation argument in favor of such agreements not valid? Such a conclusion does not appear justified. From the technical standpoint of controlling supply, effective devices to restrict and to allocate production were developed; in a few cases an industry was rationalized; the first lessons in operating buffer stocks were learned. The importance of outside producing capacity, of substitute industries, and of consumer reaction was more and more appreciated. It appears more useful to search for institutional mechanisms through which commodity agreements might become effective tools for obtaining

[46] Defenders of tin control, for example, have argued that the sheltering of high-cost lode producers is socially desirable to conserve low-cost alluvial ores. In view of a long trend of improving mining techniques (that is, probable lower future costs) the desirable way to conserve tin ores is to prevent opening up of lode mines (abandonment of which during depressions leads to wasteful depletion), and to concentrate on alluvial ores, which are less likely to be submarginal in a depression.

[47] This, at least, is true if we take a world-wide point of view. From the standpoint of national strategic interest it may be desirable for a country with low-cost deposits to favor a price level or a quota system under which foreign high-cost deposits available to potential enemies are depleted; to conserve domestic deposits by direct action is simple for a sovereign nation. In this connection it may be noted that the United States is pursuing an opposite policy in nonferrous metals, for example, copper. She is a high-cost copper producer but "protects" her copper interest by high tariffs. During the depression of the 1930's the 4 cents a pound revenue tax on copper imports was practically an embargo. A similar situation prevailed for lead and zinc. Likewise, the United States tariff is a potent factor in the depletion of domestic oil resources. From the standpoint of conserving strategic resources such a policy is questionable. More will be said about this presently.

these and other ends in the future, than to join the chorus condemning their past performance.

Commodity Agreements and Conservation: Conclusion

This brief review of the historical experience points to the conclusion that commodity agreements could become socially acceptable tools of conservation policy under two conditions: first, that their objectives be broadened considerably; second, that agreements be concluded and executed in the framework of international institutions with sufficient authority to induce realization of these objectives.

The kinds of objectives that are relevant for avoiding wasteful depletion of flow and stock resources have been indicated. The types of production control necessary to obtain these objectives do not go further in degree than did controls employed by many commodity agreements in the past; only the timing and the incidence among the various portions of resource industries would be different. Promulgation of rules about technological practices (codes) go further than provisions in the past. However, codes to avoid wasteful depletion would not be a revolutionary step. Much discussion has been directed to the problem of codes protecting the interests of consumers and workers in commodity agreements.[48] Codes of the kind suggested here are, or have been, parts of international conventions on fisheries, migratory birds, upland game, and scenic resources mentioned earlier.

The historical experience shows that commodity agreements between the major producing countries, with or without "advisory" representation of consumers, were not interested or were powerless in attacking the problems of wasteful depletion and of other socially undesirable effects of economic instability. What is needed are international institutions with sufficient authority to steer commodity agreements toward social objectives. Whether

[48] *The International Labour Code, 1939, Adopted by the International Labour Conference, 1919–1939* (Montreal: International Labour Office, 1941), 920 pp.

See also U. S. Department of State, *United Nations Conference on Food and Agriculture, Hot Springs, Va., May 18–June 3, 1943. Final Act and Section Reports* (Washington: Govt. Print. Off., 1943), Resolution XXV, pp. 3–4. (U. S. Department of State Publication 1948. Conference Series 52).

the Economic and Social Council and the functional economic organizations of the United Nations will supply this need remains to be seen.

Obviously, a mere framework of even the best international institutions does not transform a commodity agreement from what some people regard as the dangerous black wolf of social economics into a friendly white lamb. It all depends on how effective the people of the world and their governments allow these institutions to be. However, in the future, as in the past, conditions in the world markets for raw materials will bring about attempts at collective international action. A serious effort should be made to attain social objectives from an inevitable development. Avoidance of wasteful depletion is not the least of such objectives.

International Interests of a U. S. Conservation Policy

So far we have considered international tools of conservation policy without regard to special interests of individual nations. This approach is fairly realistic for the conservation of flow resources. Conservation of flow resources—in particular, realization of a safe minimum standard—benefits many nations and harms none.

Mutual benefits accrue even if flow resources in one country are conserved at the expense of depletion in another. Within limits (that is, above the safe minimum standard) and for a certain period of time, such conservation and depletion may be in the interests of both countries. An illustration of such a situation is the grain trade at the end of the nineteenth and the beginning of the twentieth centuries between central and western Europe on one side and North and South America and Russia on the other. Corn (maize), barley, wheat, and oats from the Middle West, Argentina, and the Ukraine enabled Europe to conserve its soil (through greatly expanded livestock and manure production based on imported feeds) while the virgin soils in the exporting countries were depleted. Still, such depletion was socially desirable because it enabled these countries to build up their industrial economy. Some countries (United States and Russia) are at present economically independent of grain exports; and conservation rather than depletion of soil fertility has become a socially desirable objective. Similar examples exist for forestry and grazing.

In the conservation of flow resources, conflicts between national interests are not very important even in wartime. This is not true of stock resources: in them, conservation has an immediate bearing on the military interests of individual nations. Such interests are, for example, keeping stockpiles for national emergencies, building and maintaining stand-by capacity, keeping open access to foreign resources, preventing potential enemies from access to resources, and forcing them—before actual military engagement—to deplete resources to which their access cannot be prevented. Obviously, national interests in these objectives may conflict. It is necessary, therefore, to consider such conflicts. This will now be done by focusing on the United States.

Conflicting national interests in the conservation of stock resources does not mean that our previous emphasis on international coöperation in conservation has merely academic significance: A nation, no matter how large or powerful, that pursues its own interests in the conservation of stock resources without consulting the interests of other nations will soon add to its potential enemies and weaken potential friends. A narrowly conceived international resource policy may thus defeat its own objective of military security. This holds particularly for the United States. The comparative resource position of this country, although not without weak points, is basically so strong as to permit a calm examination of national policy in a spirit of international coöperation.

Exploitation of Resource Monopolies

Little needs to be said about the possibility of using a monopoly position in stock resources for national security and economic gain. (As explained in chapter 14, exploitation of a pure monopoly tends toward conservation as compared with a situation in which pure competition prevails.) Technically such possibilities are confined to very few resources. Helium and molybdenum are the best-known examples. Even here, the monopoly position of the United States is not perfect.

Technological advance rapidly changes the military status of resources, and attempts to deprive others in peacetime will hasten these changes. Such a policy will add to international friction without doing serious damage to potential enemies.[49]

From the economic standpoint, exploitation of resource monopolies in a way that neglects the interests of importing countries has proved unprofitable in the long run for similar reasons: potential outside capacity can nearly always be developed if an appropriate economic effort is made, and it will stay in production because of high fixed and sunk costs; modern technology has increased the economic possibilities of substitution; other nations will not tolerate such exploitation for very long and will retaliate economically and politically.

Accessible and Inaccessible Resources

It is convenient for our further analysis to differentiate between two groups of stock resources, those to which the United States can maintain access in wartime and those to which such access cannot be maintained. The actual geographic location of these two groups and the strategic reasons for and the tactical means of maintaining access during a war need not be discussed in this study.

With respect to accessible resources (in the sense just indicated), there are mainly two considerations: first, the time of exhaustion of accessible resources relative to that of inaccessible ones; second, the speed and economy (in terms of manpower, equipment, transportation facilities) with which these resources can be obtained during a war.[50] The two most popular proposals —namely, freezing stock resources and stockpiling them at home —may be reviewed in the light of these considerations.

[49] Depriving "aggressor" nations of resources through *collective* sanctions is outside the field of the present inquiry. However, the statement made in the text may also be applicable to a policy of collective resource sanctions in peacetime. On this problem see P. D. Merica, "Mineral Control—Wise or Unwise?" *Mining and Metallurgy*, Vol. 25, April 1944, pp. 205–207.

[50] For example, sometime in the future the great oil pools of the New World may be exhausted. This would be of military consequence to the United States only if other nations still had safe access to oil and if other fuels (shale oil, synthetic oil from coal, alcohol from vegetable carbohydrates, atomic power) were not economical substitutes in terms of the strategically relevant factors mentioned iu the text.

Freezing and Depletion of Domestic Resources

Since the *relative* time (in the above sense) of exhaustion is what matters, the objective of military security for the United States may be achieved by increasing *in peacetime* the rate of utilization of inaccessible resources (oil in the Near East, ores on the Asiatic mainland, the East Indies). In terms of domestic and foreign political repercussions, this may be a more realistic and more economical approach than the freezing of domestic resources.

On the other hand, present widespread artificial measures that tend toward the peacetime depletion of domestic resources should be abolished. Such measures are mainly of four kinds: first, import duties on foreign resources (oil, nonferrous metals); second, subsidies for the depletion rather than conservation of domestic resources (nonferrous metals, potash, uranium); third, special favors for the depletion of domestic stock resources in the federal income tax; fourth, export subsidies through government purchases or taking over of stock at domestic prices higher than world prices, and government selling or bartering abroad (silver). These measures appear just as unfavorable for international economic stability as for the military security of the United States.

Removal of these measures, in combination with increased investment in foreign resource development and with conserving subsidies (for maintaining and increasing facilities, for exploration, and for research) rather than depleting subsidies for domestic resources, will exert a powerful, though indirect influence toward retarding domestic depletion without freezing. Freezing —for example, of domestic oil resources for future Navy use— may be looked upon as a form of stockpiling and should be judged by the same criteria as the latter. These criteria may now be considered.

Stockpiling of Accessible Resources

Stockpiling for military security of accessible resources is only one alternative. As just suggested, other alternatives are subsidies for maintaining private stand-by capacity, including subsidies for exploration and research (especially important for the oil industry and mining), or public shadow plants. Stockpiling has serious defects as compared with these alternatives.

According to experience during the last two wars, stockpiles would have to be "sufficiently" large to avoid tying up additional (that is, above peacetime requirements) productive services in resource industries. Uncertainty about this sufficiency may necessitate a diversion of manpower, equipment, and transportation facilities for the production of stock resources at the very time—during a war—when these services can least be spared. Moreover, a large stockpile may create a kind of Maginot Line mentality: the military and civilians alike may neglect exploration for new deposits and research in production and refining technology and substitution.

With respect to accessible resources (inaccessible resources will be taken up presently), military stockpiles may well be confined to the following two cases.

First, military stockpiling is sound if sufficiently large quantities of basic resources, clearly superior to possible substitutes, can be stored more cheaply than adequate stand-by capacity can be maintained in peacetime and operated in wartime. This may be true for freezing some underground deposits of oil and basic metals, such as high-grade iron ore. Such a policy is already in force to a limited extent for oil. The Truman Committee has recommended that the remains of high-grade iron ore in the Mesabi Range should be set aside as a national reserve.

Second, military stockpiles are desirable if they are limited to those quantities which are needed in the beginning of an emergency until stand-by capacity comes into full production. This applies, for example, to those nonferrous metals, high-grade deposits of which have already been depleted, which are expensive to store in large quantities aboveground because of natural deterioration, and the military status of which is in flux because of technological substitution.

It is sometimes argued by those in favor of large military stockpiles that, regardless of the type of resources involved, they may serve a double purpose: they may be used as buffer stocks in the interest of economic stability. Some doubts may be raised about the cogency of this argument.

A strictly military stockpile would be of little use as a buffer stock because the military would correctly object to liquidation when prices rise; historically, major wars have never started at the depth of depressions, but have broken out when prices were rising, partly under the influence of armaments to change the *status quo* established by the last war, and "to maintain peace."

A military stockpile, far from being useful as a buffer stock, may seriously interfere with the operation of the latter. The necessary size of the pile will be rather vague; even the best military minds could not give exact figures. This question of size will be subject to political pressure by interested groups at home and abroad. If, for example, at rising prices a buffer stock must be liquidated, producer groups may argue convincingly that the military stockpile needs to be augmented. If prices are falling, consumer interests may point to an already large military stockpile as the "cause" of falling prices and may object to building up a buffer stock at the right time. This argument may, in fact, not be entirely absurd. If the trade feels that a large military stockpile is subject to political pressures (as it will be, because of its indefinite size), such a pile may in time of political changes, domestic as well as foreign, have considerable speculative influence upon prices.

Finally, from the standpoint of economy, such a stockpile will be rather expensive and, as already suggested, may not save additional wartime investment in production.

Inaccessible Resources

We may now turn to inaccessible resources in the above sense. These are less numerous and important than the group just discussed. With respect to them, one recommendation has been a program of military stockpiles combined with high peacetime import duties, or even import prohibition in order to force the development of domestic substitutes.

This approach, used by Germany before the last war, imposes

a great burden upon the peacetime economy, unless economical substitutes can be developed in time. In the latter event this approach is comparable to an educational tariff. If large quantities are needed which are expensive to stockpile (natural rubber), the only alternative is to develop substitute industries.

For the United States, however, most inaccessible stock resources are vital for national security only in rather small quantities (e.g. steel alloys, industrial diamonds, mica, quicksilver); sufficient stockpiles are physically and economically no problem. Political pressure groups are not interested, since no domestic producers are involved, and since the price of these resources is of small consequence for the final product.

Except for more emphasis on military stockpiles, the same policies may be used that were suggested for accessible resources. Removal of import duties and increased investment in foreign resource development will keep domestic prices low. This situation in itself will create a stockpile in the form of regular trade inventories and scrap. A policy of low domestic prices will tend to push resources into less essential uses. In an emergency, these resources can be transferred to strategically vital purposes. Such a policy, however, should be supplemented by subsidies for research and for stand-by capacity in substitutes.

Development of Foreign Resources

The foregoing discussion points to the desirability of United States investment in the development of foreign resources, especially accessible ones, in order to conserve domestic resources. A few problems of such a development may be taken up in the remainder of this chapter because they have an important bearing on resource conservation internationally.

An open-door policy in foreign investment has been one basic principle of United States foreign policy. However, more is required than the open door. No less important in the long run are the interests of the countries within whose boundaries accessible stock resources are found.

Sometimes the whole economic and political development of such countries depends on how one or a few stock resources are developed by foreign capital. Oil in Venezuela, tin in Bolivia, copper and nitrates in Chile are examples. The enlightened self-interest of the United States requires that rate and methods of depletion be related to the development of local economies, and that the resulting foreign exchange be employed for that purpose. Otherwise, whole countries may permanently be left in a condition which in the United States has temporarily occurred in relatively unimportant areas when abandoned mines and cutover forests left ghost towns and stranded people.

In this respect, the past record of foreign investments in underdeveloped countries has not always been encouraging. The absence of good will and of economic contentment among local populations may vitally concern the capital-exporting countries themselves from political, military, and public-health standpoints. This was vividly illustrated by certain United States, British, French, and Dutch experiences in the Pacific and Far Eastern theaters during the last war. Since the repercussions from shortsighted depletion are world wide, the United States is interested in preventing them, even if its own citizens are not directly concerned as investors. The United States cannot wash its hands of colonial exploitation or politically enforced terms of trade, even when other capital-exporting countries are primarily involved.

Foreign Resource Development and Population Growth

One frequent result of the penetration of Western civilization into economies whose resources are to be developed is an increase in population. As we know, such an increase is caused mainly through a decrease of death rates, especially in early childhood. Only industrialization—not merely a mechanization of raw-material production and agriculture—can absorb this population increase. Experience has shown that in industrial employment and urban surroundings a non-Malthusian balance between birth and death rates—through decrease of birth rates—is reached much faster than in agricultural employment. A mere mechanization of raw-material production aggravates the problem. Such mechanization tends to create an underemployed coolie proletariat, already much in evidence in plantation areas (Malaya, the East Indies, the Caribbean). It appears necessary, therefore, economically as well as politically, to install local refining and manufacturing capacity and secondary industries, instead of following the attempt of some colonial powers to confine underdeveloped countries to the first stage of industrialization.

The objection may be raised that industrialization is too slow a process in obtaining a non-Malthusian balance between birth and death rates. If such an objection is to be meaningful, the word "slow" must be defined in terms of some alternative. To many students, this alternative is birth control. There is some doubt whether efforts in this direction will be quickly effective if children are still an asset to individual families—even though they may be regarded as a liability to a population as a whole. In agriculture, children are an asset to individual families; in industrial occupations and urban residence, children are a burden—especially if women are employed outside the home. At the same time, such an occupational and residential change favors a shift from habit patterns to calculativeness in matters of family size. Thus, transformation of agricultural economies into industrial ones would seem a prerequisite in efforts to spread birth control—and not the other way around.

Population Increase and Conservation of Resources

Birth control has been advocated directly as an international tool of conservation policy. In the opinion of the neo-Malthusians referred to in chapter 1, depletion of soils, forests, and grasslands is largely caused by the "pressure" of people upon the land. They propose, therefore, checks on population increase as a tool of resource conservation. With respect to underdeveloped countries, some of them even go so far as to suggest that medical and nutritional assistance be withheld by the United States—for example, in a Marshall plan for Asia—in order to keep the Malthusian checks operative or to force native governments to provide education and facilities in voluntary birth control.

Some doubt has already been raised with respect to the practicability of these proposals if an industrialization of the economy is not brought about first. Their moral and ethical side is outside the scope of this study. This study is, however, directly concerned with the relation of population pressure to resource conservation. With respect to this relation, two observations are pertinent.

Depletion of soils, forests, grasslands, and wildlife on a large scale and of serious degree can be observed in areas of the world where population pressure relative to that in other areas is very slight. This is true whether population pressure is measured by some kind of man–land ratio, or in terms of death rates, life span, or real income per person. Examples of such areas are Australia and the United States.

On the other hand, in many areas of the world, resource conservation has been practiced for centuries in spite of a high population pressure. Examples are the Inca Empire of ancient Peru, medieval Europe, present-day Japan, and western China (Szechwan).

It is apparent, therefore, that a high population pressure is neither a necessary nor a sufficient condition for resource depletion. Too much emphasis on population pressure tends to confuse issues and to retard public awareness and acceptance of those educational and institutional changes that are needed.

Educational and Institutional Changes

The changes that are needed are closely linked to the educational level of the native population and to their social institutions. People must learn to utilize their resources, and institutional obstacles to resource development must be overcome.

Institutional obstacles must be removed by the people themselves. Outsiders may help if they are asked to do so by native reform groups. But unsolicited advice or pressure from the outside only creates resentment and opposition.[51] To help other people in understanding American institutions is one thing, to

[51] Obviously, any reform will create resentment somewhere. Sometimes it may be politically advantageous for native governments—even though they have asked for assistance from the outside—to direct such unavoidable resentment away from themselves toward outsiders.

insist that they accept such institutions for themselves is quite another. Even if transplantation of American institutions were possible through direct and indirect pressure, they may be entirely unsuited to the new environment without modification. This appears true particularly for the part played by governments in economic affairs. Understanding the foreign environment and respecting foreign people as equals is the first step in helping them to bring about educational and institutional changes.

The most economical way for underdeveloped countries to import foreign knowledge is through securing experts for a limited period. To be most useful, such experts should not only work on individual projects but also act as teachers for native personnel. The projects themselves should not be confined to technology but should include the field of social institutions such as taxation, tenure, and credit. Generally, the experts will want to transfer a considerable part of their earnings into their home currency. This may create difficulties for the underdeveloped country. An effective form of assistance from the United States, therefore, would be to pay the foreign-exchange portion of the salaries of experts loaned to underdeveloped countries.

Another, but possibly more expensive, way for underdeveloped countries to import foreign knowledge is through sending abroad native personnel for training. Generally, it will be more effective to send individuals who have already received some training and experience at home (in the great majority of underdeveloped countries, the beginnings of higher education are present). These individuals are more mature, better able to pick out abroad what is needed, and less likely to become permanent emigrants. Such people are also better suited for in-training in offices, laboratories, factories, and farms. In all these respects, the sending of young students is less desirable. Assistance to the underdeveloped country may well be given by the United States through financing the foreign-exchange portion of such a training program. Economically, and politically also, a large-scale program of fellowships given by the United States to foreign managers, civil servants, professional men, and skilled workers would seem a comparatively effective use of United States funds—which, after all, are not unlimited.

Foreigners who have acquired firsthand and intimate understanding of American economic and social institutions are a ferment to change institutions at home and are a far better cement for building international friendship and loyalty than interest and amortization payments on loans or gratitude for material gifts under whatever name or conditions such gifts may be granted.

It may be objected that a large-scale in-training program for foreign personnel will not be possible because of the opposition of American labor unions. The wholehearted coöperation of unions in such a program would certainly be necessary. In this respect, however, one need not be too pessimistic. There are indications that the American labor movement is becoming increasingly alert to its responsibilities in foreign relations.

Given adequately trained native managers, civil servants, professional men, and skilled workers, the possibilities of internal capital formation should not be underestimated. Even in the most underdeveloped countries, there is usually considerable saving which at present goes into private hoards, into the foreign investments of a small group of the well-to-do, and into public construction for ceremonial purposes. There is also a good deal of conspicuous consumption by individuals and governments which could be transformed into savings. Mobilization of actual and potential internal savings for capital formation, through reform of taxation and property institutions, would, in many cases, give a good start toward industrialization.

In the beginning of industrialization, imports of capital goods will be necessary. A transfer problem will exist. However, if a domestic basis for industrialization is prepared in terms of skills, institutions, and capital formation, private credit or direct investment by foreign enterprises can in normal times be secured without difficulty. On the other hand, in times of high political tension private credit or direct investment may not be available to underdeveloped countries because of their exposure to direct or indirect attack by a third country. In such times provision of government funds may be vital for the interests of the United States.

A gradual program of industrialization may seem slow. Historically, however, this was the way in which Western countries and Japan built up their industries. The ultimate objective—an increase of real income per person—may be reached possibly not later and certainly more cheaply and safely than through a forced program of government-financed exports of American equipment and American technicians to operate it. The latter program, however, is justified if the United States security demands a rapid acceleration of raw-material imports.

CHAPTER

21

COÖRDINATION OF CONSERVATION POLICY

Limitation of Our Objectives

There would be little point in trying to explain here the *need* for coördination of conservation policy, both domestic and international, in the United States.[1] So much has been written lately on this point that the duplications, inconsistencies, conflicts, and administrative inefficiencies in the present situation are fairly well known.

Many proposals have been made to remedy this situation. The most important are those of the Hoover Commission.[2] The report of the President's Water Resources Policy Commission makes some proposals for administrative coördination in the field of

[1] As noted before (chap. 2), this chapter deals exclusively with the situation in the United States.

[2] These are largely embodied in the recommendations regarding the Departments of Agriculture and Interior, based on the Task Force reports on "Agricultural Activities," "Natural Resources," and "Water Resources Projects." See U. S. Commission on Organization of the Executive Branch of the Government, *The Hoover Commission Report on Organization of the Executive Branch of the Government.* (New York: McGraw-Hill Book Co., 1949), 524 pp.

water resources.[3] Numerous individual writers have come forth with their own proposals.[4]

But these proposals, though all recognize the need for coördination, are conflicting and controversial. Some of them may create difficulties no smaller than the existing ones. Is there a basis for choosing among them?

Our objective is not to review or to criticize these proposals in detail. Neither is it to present another blueprint for government reorganization. Rather let us examine, on the basis of the preceding chapters, the major *requirements* for effective coördination of conservation policy and, from these, draw certain conclusions with respect to the major *directions* in which improvements may be sought. We may also consider how far recent proposals follow these directions.

Major Requirements for Coördination

The major requirements for coördination are created, first, by the substance of conservation policy itself, and, second, by the constitutional framework within which conservation policy is formulated and executed. The former requirements arise in part from the objectives and tools of conservation policy, and in part from the interrelations between the resources with which it deals.

The Requirement for Over-All Coördination

In the preceding chapters an attempt was made to show that conservation policy is broad in its objectives and varied in the tools at its disposal. Its broad objectives link it with the objectives of other economic and social policies. Its varied tools demand coördination not only within conservation policy itself but

---

[3] U. S. President's Water Resources Policy Commission, *A Water Policy for the American People* (Washington: Govt. Print. Off., 1950), Vol. 1.

[4] For example:

M. W. Watkins, "Scarce Raw Materials: An Analysis and a Proposal," *The American Economic Review*, Vol. 34, No. 2, June 1944, pp. 227–260.

J. W. Finch, "Conservation," *Mining Congress Journal*, Vol. 23, No. 12, December 1937, pp. 32–33.

C. H. Hammar, "Society and Conservation," *Journal of Farm Economics*, Vol. 24, No. 1, February 1942, pp. 109–123.

also with the tools of other policies. Indirect tools of conservation policy—operating through changes of interest rates, allowance for uncertainty, prices, tenure, taxation, and the like—are basically designed for economic objectives other than those of conservation (chapters 7 to 15). Even the direct tools—education, and especially subsidies, zoning, and regulations—may have far-reaching effects on the economy in addition to their effects on conservation (chapter 19). Some tools of conservation policy operate in the field of international relations (chapter 20).

Thus many economic policies have a bearing on conservation. On the other hand, problems other than those of conservation are involved in such policies. Conservation is only one phase of economic policy. It is the contention of our study that this phase needs more attention. But it would be undesirable to subordinate all economic policies which affect conservation to the latter's requirements.

The broad and variable substance of conservation policy suggests caution with respect to the proposals of those who want to satisfy the need for coördination by centralizing all conservation policy in a single agency—for example, a conservation commission or a federal department of natural resources. If such an agency were to administer all the policies that bear directly or indirectly upon conservation, most federal executive departments would have to be subordinated to it, at least in many of their functions. Such an all-comprehensive and all-powerful agency would become an administrative monstrosity.

On the other hand, the broad and variable substance of conservation policy stresses the need for an over-all coördination—that is, a coördination of all economic policies, including those dealing with conservation. How this requirement could be satisfied will be taken up in more detail below.

Conservation Policy and Interrelations of Resources

There is another aspect of the substance of conservation policy which was stressed in the preceding chapters and which is relevant here: resources are interrelated. The complementary and competitive relations in supply and demand are especially numerous and complex for flow resources. Furthermore, the most important flow resources have several common characteristics which are highly important for conservation policy. One of these characteristics was discussed under the labels "critical zone" and "safe minimum standard of conservation." Other common characteristics are the great importance of institutional factors and of habit patterns and the relatively (as compared with stock resources) minor importance of technological change and monopolistic conditions.

Thus one basic requirement for the coördination of conservation policy is that the major flow resources be treated as one complex but inseparable whole. Proposals which aim to coördinate conservation policy, but fail to take account of the close interdependence of flow resources and assign responsibility for them to separate major governmental units, cannot be regarded as improvements.

The complementary and competitive relations between the major stock resources, the great importance of technological change and of monopolistic conditions, the significance of international and military-security aspects, and the relatively (as compared with the major flow resources) minor importance of habit patterns, were repeatedly emphasized. For conservation policy, stock resources form an interrelated complex quite different from that of flow resources.

There is one important resource which does not fit clearly into either group. This resource is hydroelectric power. Hydroelectric power is a flow resource, and its planning and development are closely related to those of other flow resources, such as fisheries, recreation, and water for domestic use and irrigation. But the use of hydroelectric power has important complementary and competitive relations to many stock resources, both in supply and in demand. Such stock resources are, for example, coal, oil, uranium, and nonferrous metals—especially aluminum. After hydroelectric power has been developed—that is, after dams are built and generating capacity installed in a river system—the problems of integrated operation, distribution, and use are more closely related to those of mineral resources and of the industrial and defense sectors of the economy than to those of flow resources and agriculture.

Consolidation within Two Major Spheres

Thus coördination, both in the legislative and in the executive branches of government, may well proceed within two major spheres, those of stock and of flow resources.

Tendencies for a consolidation of conservation policy within two spheres exist in the legislatures and executive offices of federal and state governments, as discussed in more detail later. In the highly important committee system of many legislatures, frequently two committees deal with flow and stock resources respectively. In the executive branches of many state governments, two departments deal with natural resources; one is usually called "Agriculture," the other variously "Natural Resources," "Mineral Resources," and so on.

At present, such a two-sphered consolidation of conservation policy is incomplete because of historical accident in government organization, vested interests of established agencies, and public apathy. One need for coördination stems from this incompleteness —that is, from the nonfunctional separation of government responsibilities for policies which in substance are closely interrelated.

Consolidation of conservation policy within two major spheres cannot be expected to solve all problems of coördination. For example, we will see later that conflicts exist not only between major executive departments, but also between bureaus of the same department. By and large, however, a common secretary is a strong coördinating force.

The Constitutional Framework of Conservation Policy

Any discussion of coördination of public policy in the United States must be realistic in terms of the constitutional framework within which government is organized. Two aspects of this framework are especially relevant here.

First, the Constitution divides government responsibility in the resource field between federal and state authorities. The portion of responsibility assigned to the states is particularly important. This holds both for indirect tools of conservation policy (such as taxation, tenure, and credit) and for direct ones (such as zoning and regulation of practices). With respect to international tools, federal authority is paramount. But even here, state participation is essential—for example, in policies affecting coastal and anadromous fisheries, and in enforcing federal regulations concerning migratory birds.

Most proposals for the coördination of conservation policy concern only the federal level. Such coördination would leave an important part of conservation policy untouched. Even if it were desirable—which may be doubted—it would be difficult constitutionally to endow the federal government with the necessary authority over the states. Thus, the constitutional framework poses the problems of federal–state and interstate coöperation in conservation policy.

Second, the Constitution divides federal responsibility in the resource field—as in most others—between the legislative, the executive, and the judicial branches. State constitutions have similar provisions.

Most proposals for the coördination of conservation policy concern only the executive branch. Even if such proposals were effective and should actually be realized, coördination on the legislative level would still be a problem.

At present, the legislative mandate for conservation policies is contained in many unrelated federal and state acts, international treaties, state compacts, and local ordinances. Many of these mandates contain very detailed instructions to the executive with respect to administrative organization. Proposals have been made that a conservation act should contain no more than general principles, and that broad powers of policy formulation should be delegated from the legislative branch of government to a department of conservation or to a conservation commission in the executive branch.

In specific circumstances the Congress has delegated legislative powers to some executive agencies—for example, to the Department of Agriculture and to the independent regulatory commissions (Interstate Commerce Commission, Federal Trade Commission, Federal Power Commission). Such delegation has been held by the Supreme Court not to conflict with the threefold separation of governmental powers required by the Constitution. The Supreme Court has conveniently defined such powers as quasi-legislative and quasi-judicial.[5] But even if it were desirable —and again this is open to doubt—it appears unlikely that such broad powers as would be necessary for conservation policy could constitutionally be delegated to an executive department or commission.

Thus the constitutional separation of government power poses the problem of *legislative* and not merely *executive* coördination of conservation policy.

Conclusions: Directions for Improving Coördination

From the foregoing points we may conclude that better coördination of conservation policy requires improvement in the following four directions:

1. By establishing or strengthening over-all planning and reviewing agencies in the executive branches of federal and state governments.
2. By consolidating executive activities concerning flow and stock resources in two departments.
3. By consolidating legislative work on flow and stock resources in two legislative committees.
4. By strengthening federal–state and interstate coöperation.

Possibilities for improvements in these four directions may now be taken up in turn.

A Federal Planning and Reviewing Agency

The disappearance of the National Resources Planning Board in 1943 ended one attempt to establish an over-all planning and reviewing agency in the federal executive branch. However, a previously noticeable trend toward greater prominence of the

[5] Robert E. Cushman, *The Independent Regulatory Commissions* (New York: Oxford University Press, 1941), 780 pp.

Bureau of the Budget as general staff agency of the federal executive was strengthened.[6]

There is little doubt that the Executive Office of the President is the proper place for the agency which we have in view. It is less obvious whether such an agency should be a division in the Bureau of the Budget or, instead, a bureau of planning and review coordinate with the Budget Bureau. (Essentially, a bureau of planning and review is proposed both by the Hoover Commission and the President's Water Resources Policy Commission, but such proposals are confined to water resources—which seems insufficient for reasons already indicated.)

For the latter arrangement, the argument is that the important day-to-day and year-to-year functions of the existing major divisions of the Budget Bureau may overshadow and retard the growth and influence of a new division.

On the other hand, no planning and review within the executive branch would be of much consequence without the support of the chief executive and his general manager, the director of the budget. Given this support, a properly staffed planning and review division within the Bureau of the Budget would gain from close association with the other divisions, and vice versa. Through the budget lever, such a division could exercise more power to coördinate executive agencies than could a separate bureau. Judging from experience with the National Resources Planning Board, a division subordinated to the director of the budget would probably be more acceptable to Congress than a coördinate bureau. Establishment of such a division appears, therefore, the most appropriate solution for over-all coördination of economic policy in the executive branch. At present the facilities of the Budget Bureau are not sufficient for this task.

---

[6] Under President Franklin D. Roosevelt the sphere of responsibility of the Bureau of the Budget (established by the Budget and Accounting Act of 1921) increased greatly. In 1937 the President's Committee on Administrative Management recommended that the Bureau of the Budget assume the role of the President's agent in coördinating policies of the executive branch and in unifying departmental operations. Following these recommendations, the Reorganization Act of 1939 transferred the bureau from the Treasury Department to the Executive Office of the President. The duties of the bureau were subsequently redefined and enlarged by executive orders.

The planning and review division of the Bureau of the Budget needs sections dealing with flow resources, mineral resources and power, fiscal and monetary policies, labor and public welfare, and other important aspects of economic policy. The two resource sections need staffing with competent specialists, able to deal both with the technology and economics of conservation. So wide are the ramifications of the conservation problem, however, that the permanent staff could not well include experts equipped to deal with all classes of problems. The planning and review division, more than any executive agency, might utilize temporary consultants from within and without the government, together with research facilities in government departments, universities, and other public institutions.

The planning and review division and particularly its two resource sections might well maintain small regional staffs of a caliber that would command the respect of federal and state agencies. This proposal would involve no revolutionary changes in the organization of the Bureau of the Budget. At present a special division, the Field Service, helps that bureau to investigate the operation of federal departments in the various regions of the country, counsels with federal agencies with a view to improving coördination within regions, and consults with state and local officials regarding federal policies. For reasons already indicated, federal–state coöperation appears especially important in conservation policies. On this point, more will be said below.

A Nonadministrative Planning Commission?

A National Planning Commission has been suggested as a center where plans of the executive branch, of Congress, and of organized industry, labor, agriculture, and the like, could be given the spotlight for public information and discussion, independent of administrative functions. Such a commission would serve primarily as a means of publicizing the plans of other agencies, in contrast to the anonymous operation of the planning and review division suggested above. To that end it would have only a small staff to help organize hearings and digest the materials presented to it. It would farm out research projects to government and private agencies as did, for example, the Temporary National Economic Committee.

Such a commission may be useful. It would not, however, obviate the need for strengthening the coördinating machinery along the lines just discussed. For the coördination of conservation policies, reforms in the latter direction appear more important.

Self-Coördination by Federal Agencies?

It would not be fair to overlook attempts by federal agencies to coördinate themselves. There have been several such attempts.

Coördination of bureaus within the same department may appear to have a simple administrative solution: through the common secretary. However, it is not so simple if the activities of each bureau can marshall political support for different reasons or from different groups of voters. In such situations, the secretary concerned may find it inconvenient or difficult to insist on an effective coördination.

An example in the Department of Agriculture is the coördination of the Agricultural Conservation Program of the Production and Marketing Administration (previously Agricultural Adjustment Administration) with the programs of the Soil Conservation Service and the Forest Service. Inconsistencies in these programs have existed on the state and county level for many years. Attempts to consolidate the two former programs through legislation have failed. Only in the spring of 1951 did the secretary of agriculture move to coördinate state and local planning of these programs: [7] the programs are to be determined "jointly." However, the administrative organization of all agencies remains unchanged; their objectives [8] and tools [9] have been inconsistent in the past; and the local groups on which support rests—the state

[7] U. S. Department of Agriculture, Office of the Secretary, *Coordination of the Department's Agricultural Resources Conservation Services* (Memorandum No. 1278, Office of the Secretary, U. S. Department of Agriculture, February 15, 1951).

[8] Short-run for the A.C.P., long-run for the S.C.S.

[9] Payments for land clearing, drainage, and reducing acreage of depleting crops (chap. 9) in the A.C.P., integrated "farm plan" for soil and water conservation in the S.C.S.

and county P.M.A. committees and the soil-conservation-district directors, respectively—remain unconsolidated. One will have to defer judgment on whether "joint determination" will mean effective coördination of planning and execution.

In the Department of the Interior, conflicts exist between the Bureau of Reclamation on one hand and the Fish and Wildlife Service and the Park Service on the other. Here, also, political support comes from different groups of organized interests—the various reclamation associations on one side and the many clubs and associations of recreationists and sportsmen on the other. The present conflicts concerning the Glacier National Park, the Dinosaur National Monument, and the Tule Lake Wildlife Refuge are examples. The secretary of the interior, confronted with opposing advice and pressure, is having difficulties in reconciling them.

With respect to coördination of different departments—rather than bureaus within the same department—the Departments of Agriculture and Interior have each appointed a land-use coördinator. These "coördinators" may inform, discuss, and suggest, but they have no authority over their own departments.

More important is the Federal Inter-Agency River Basin Committee in Washington with its regional bodies for the Missouri, Arkansas, and Columbia basins and its subcommittee dealing with power problems in the Southeast.[10] Some students believe that the Inter-Agency Committee has been "completely ineffective." [11] Although this may be somewhat too harsh a verdict, there is danger that government by the Inter-Agency River Basin Committee may lead to interdepartmental compromises on individual projects on a *quid pro quo* basis but not to true coördination of

---

[10] By the 1938 Flood Control Act both the Department of the Interior and the Corps of Engineers were authorized to do research in flood control and related problems. As a result (and possibly under the influence of demands for the extension of integrated river-basin authorities modeled after the TVA) the two agencies joined with the Department of Agriculture in a voluntary tripartite agreement in August, 1939, for the exchange of information and reports. The agreement was extended to the Federal Power Commission in December, 1943, and to the Department of Commerce in September, 1946.

[11] This, for example, is the opinion expressed in one minority report of the Hoover Commission.

planning. In some instances the compromises may be reached at the expense of the taxpayer and may make it difficult for outsiders to assess the weaknesses of a proposed "unified" plan.[12]

Mention may also be made of regional councils of federal agencies—for example, the Pacific Coast Federal Regional Council. The purpose of these councils is discussion and coöperation rather than coördination among the regional offices of federal departments and bureaus.

The National Security Resources Board is an attempt to coordinate, at the cabinet level, the regular federal departments in their activities relating to the present defense effort. Members of the board are the secretaries of the State, Defense, Treasury, Interior, Agriculture, Commerce, and Labor departments under a chairman appointed by and representing the President. Obviously, it is very difficult to appraise the effectiveness of the board from the outside. Several changes of the chairman, limitations on his authority, and existence of other special defense agencies with wide powers are unfavorable factors. In any event, the specific purpose of the board does not make it a suitable substitute for a permanent, over-all planning and reviewing agency in the federal executive.

The participants themselves in these and other attempts at self-coördination would possibly agree that the results were not sufficient. It is doubtful whether it is possible for coördinate agencies to coördinate themselves at all effectively. In contrast, the proposed division of planning and review in the Budget Bureau

---

[12] The task force on natural resources of the Hoover Commission has the following to say on this problem: "The development agencies sometimes compromise their differences. After sharp clashes over plans for the development of the Missouri Basin, the Corps and the Bureau announced complete agreement on the Pick-Sloan plan. Analysis of that plan reveals the fact that it contains many projects which previously had been subjected to devastating criticism by one or the other agency. The 'compromise' consisted for the most part in a division of projects, each agency agreeing to forego the privilege of criticizing projects assigned by the agreement to the other. The result is in no sense an integrated development plan for the Basin, and there is serious question in this case whether agreement between the two agencies is not more costly to the public than disagreement."—*Task Force Report on Natural Resources [Appendix L]: Organization and Policy in the Field of Natural Resources. A Report with Recommendations*, prepared for The Commission on Organization of the Executive Branch of the Government. (Washington: Govt. Print. Off., January 1949), p. 24.

would make itself felt in very short order, even if the budget lever is merely kept as a "force in being."

A Planning and Reviewing Agency in State Governments

In most states, machinery for planning and review in the executive branch is even weaker than in the federal government. Improvements might be made along the lines suggested for the federal executive. Many states do not possess a counterpart of the Budget Bureau. In California, for example, the state budget is formulated by the executive agencies in coöperation with the Division of Budgets and Accounts of the Department of Finance; this department administers the budget after enactment of the budget bill by the legislature. Such coöperation and administration, however, is largely from the point of view of finance and accounting rather than of coördinating executive activities.

During the 1930's many states created planning boards with the assistance of the National Resources Planning Board. However, the state planning boards were more concerned with specific public-work schemes than with continuing administrative coördination. Few of them succeeded in recruiting qualified personnel.

Toward the end of the Second World War, several states appointed commissions for "postwar" planning; but these agencies have not been placed in a strategic position comparable with that of the suggested planning and review division in the federal Bureau of the Budget.

Consolidation of Federal Executive Activities

Why consolidation of conservation policy in two major resource departments is advisable was indicated earlier. The names "agriculture" and "mineral resources and power" may be suggested for them. However, the particular names chosen are immaterial.

The proposed department of mineral resources and power would include such presently independent agencies as the Bonneville Power Administration and Tennessee Valley Authority. Arguments are fairly evenly divided on whether hydroelectric power should be grouped with flow or with stock resources. As implied earlier, it could best be grouped with other flow resources during the planning and development (construction) phase, but with stock resources after development—the administrative phase. Because of the more permanent character of the latter phase, a grouping with stock resources is suggested here. However, the requirement for over-all coördination, already stressed, is especially important for hydroelectric power.

We may now consider some of the obstacles and objections to such a consolidation in the executive branch.

Political Obstacles to Consolidation

The obstacles to such a consolidation are largely political—that is, are due to the strong support of some agencies from certain groups of resource users and the Congress.

It would be a mistake, however, simply to explain this support by partisan interests and by the "pork barrel" argument—as is frequently done in the literature. Let us take a well-known example: the political obstacles to the transfer of the water-resources activities of the Department of Defense (Army) to the Department of the Interior or of Agriculture. Admittedly, the Army Engineers are the favorites of some pressure groups and of some members of Congress. However, the "pork barrel" is not the whole story.

The Army Engineers are mainly builders of a project. This leaves room for state and other local participation in the administration of a project after it has been built. By and large such local participation—which, of course, should extend also to the financial responsibilities—is desirable from the standpoint of economy and administrative efficiency. The Bureau of Reclamation, in contrast, is in itself as much an administrative as a construction agency. This is due partly to the reclamation laws and partly to interpretation of the laws under the influence of traditions and personalities.

A detailed discussion of administration versus construction by federal agencies is outside our present interest. However, those who (as the writer) favor a transfer of the water-resources activities out of the Department of Defense (Army) should ponder whether the strong political position of the Army Engineers is solely due to the "pork barrel," or in part to other features of their activities which make for a better institutional fit.

The Hoover Commission's Proposal

Although many students may agree that executive resource activities should be consolidated in two major departments, they disagree about the division of responsibility. An example is the report of the Hoover Commission.

The majority of the Hoover Commission assigns water development, wildlife management, and recreational fisheries to the Department of the Interior (Natural Resources); forest management, range management, and irrigation management to the Department of Agriculture; and commercial fisheries to the Department of Commerce. This report is frequently not in agreement with the recommendations of its own task forces.

It is difficult to see how existing conflicts between the Departments of the Interior and Agriculture could be reduced by such a reorganization. In addition, new conflicts will arise in fisheries. How far, for example, is the Pacific salmon a "recreational" resource and therefore to be administered by the Department of the Interior, and how far is it a "commercial" resource and therefore to be administered by the Department of Commerce?

It is not surprising that the majority report of the Hoover Commission (signed, with numerous objections and abstentions regarding individual recommendations, by only seven of the twelve commissioners) is opposed by two minority reports, one of which (signed by the vice-chairman and two commissioners) is in many respects stronger and more consistent than the majority report. The strength of this minority report is the recognition that water, wildlife, forest, and range are so closely interrelated both in planning and administering their development that they should be the concern of one executive department.

This position is in agreement with that taken in this study. However, this minority report would like to split the executive concern for these same resources along the lines of administration of publicly owned resources and service to privately owned resources. The former function is assigned to the Department of the Interior (Natural Resources) and the latter to the Department of Agriculture. On this basis the administration of water, forests, range, and wildlife is delegated to the Department of the Interior as far as these resources are in public ownership. The same resources are administered by the Department of Agriculture if they are in private ownership.

It may be submitted that such a split in administration will lead to no less serious conflicts, duplication, and waste than the existing situation. The technical problems of resource development are not changed by ownership. As was brought out earlier, public concern is just as much with private as with public resources. Conservation policy as understood in this study is *mainly* concerned with privately owned resources. Privately and publicly owned resources of the same kind are closely interrelated in use (forests, range, wildlife).

The idea that the present Department of Agriculture is merely a service agency (research and extension) for farmers is obsolete. The planning and regulatory functions of the Department of Agriculture have become increasingly more important. The U. S. Forest Service has operated as a bureau of the Department of Agriculture for many years. This service is largely concerned with the administration of publicly owned resources. Its record in upholding the public interest against pressure groups and in recruiting a public-spirited personnel would stand comparison with the record of any other government agency.

In addition, a split in administration between public and private ownership might hamper the proper choice of the tools of conservation policy—for example, if a change from regulation to public ownership meant a transfer of administration from one department to another.

The other minority report (endorsed by two commissioners) need not be considered here because it favors the *status quo*. In particular, it seeks to continue the present split of water development between the Departments of Defense (Army), Interior, and Agriculture.

Consolidation of executive concern with all flow resources (except hydroelectric power) in a department of agriculture, and of that with all stock resources and power in a department of mineral resources and power, would realize the aims of both the majority and one minority report of the Hoover Commission.

Consolidation and the Department of Agriculture

The proposal made in this study could be criticized because the already big Department of Agriculture would be made still larger by addition of construction functions in the field of water resources.

A reorganization of the Department of Agriculture is long overdue. Some of its present banking, trading, and inspection functions could well be transferred to other existing agencies.

Most, if not all, construction can be handled by private commercial enterprises on the basis of competitive bidding. For planning, obviously, a good permanent staff of engineers is needed. But over-all economies in the field of water-resources development would be possible if the present separate surveys, plans, and construction activities by the Army Engineers, the Bureau of Reclamation, and the Department of Agriculture could be consolidated.

Integration, but in general no decrease in the number, of functional specialists in Washington, in regional headquarters, and in federal experiment stations would be possible. At present, several bureaus of the Departments of the Interior and Agriculture maintain separate staffs of functional specialists concerned with soils, water, range, and forests. These specialists have the same training and deal with the same problems in the same general area. Sometimes, an individual resource user (irrigation farmer, cattleman, recreationist) or different resource users in identical situations are confronted with conflicting policies and advice emanating from two or more bureaus and departments. Integration would reduce such confusion and increase the services rendered.

Likewise, local administrative personnel of various agencies should be integrated, but, in general, should not be decreased.

This may make it possible to reduce the size of administrative units—for example, ranger districts.

A Department of Mineral Resources and Power

The proposal made in this study may also be criticized because mineral resources and power may not appear important enough to warrant a separate department with an officer of cabinet rank at the head.

But is such a department unwarranted for an industrial country in the atomic age? The development of mineral and power resources, their distribution, their far-reaching international aspects, the provision of stand-by capacity for emergencies, the development of substitutes, the payment of subsidies, and the economic integration of atomic with other forms of energy, are some of the problems—among many related ones—with which a department of mineral resources and power would be concerned.

These problems rank among the most vital and most complex issues confronting this country. At present, the executive concern for them is scattered among several agencies. Weaknesses in their coördination may become more important for national survival than the similar situation in the field of water resources. Yet the main attention of administrative reformers and of the public in general is focused on the latter situation.[13]

Consolidation of State Executive Activities

In the states, consolidation of executive activities in the resource field in two departments is sometimes further advanced than in the federal government. However, in some states three or more departments are concerned with flow resources alone.

In California, for example, the Division of Water Resources is in the Department of Public Works.[14] Parks, forests, and recre-

---

[13] To avoid misunderstanding: there is no intention to suggest that the independent character of the Atomic Energy Commission be changed. There are many ways in which a strong Department of Mineral Resources and Power can aid the specific purposes which are the proper domain of the A.E.C. and other agencies mainly concerned with national defense.

[14] In water resources, the State Water Resources Board (which in its actual operation relies on the Division of Water Resources) fulfills some planning and coördinating functions.

ation are the concern of the Department of Natural Resources. Fish and game, until recently administered by that department, are now the responsibility of a special department. Farming and ranching are under the Department of Agriculture. In this state, conflicts between the conservation of fisheries (salmon, steelhead, trout) and waterfowl resources on one side, and water development for irrigation and power (dam construction, water diversions, low priority of water for resting and wintering areas for waterfowl) on the other, are especially serious—for example, dam construction may destroy anadromous fisheries; lack of water allocated for resting and wintering areas may critically affect the whole Pacific Flyway.

Coördination in the Federal Legislative Branch

Better coördination of conservation policy in the executive branch of government does not appear sufficient. A general conservation act delegating policy formulation to the executive and eliminating the need for legislative coördination would not be suitable for the purposes here under consideration. Such an act would not be constitutionally feasible or politically wise from the standpoint of the checks and balances so essential for democratic government. It appears probable that conservation policy will continue to be formulated through numerous individual acts —and politically desirable that it should be so formulated. This situation poses the problem of legislative coördination.

The legislative branch, like the executive, displays certain trends that may be utilized for our purpose. Within and without Congress, there is a strong desire to modernize partly obsolete legislative machinery. The deep and growing interest in Congressional self-improvement is shown by the introduction of more than fifty legislative proposals for specific reforms.[15] Some of these

---

[15] For details see *The Reorganization of Congress, a Report of the Committee on Congress of the American Political Science Association* (Washington: Public Affairs Press, 1945), 89 pp.

See also:

Lindsay Rogers, "The Staffing of Congress," *Political Science Quarterly*, Vol. 56, No. 1, March 1941, pp. 1–22.

Arthur W. Macmahon, "Congressional Oversight of Administration: The Power

proposals have been realized through the Legislative Reorganization Act of 1946.[16]

Two aspects of Congressional reorganization are of special interest for the coördination of conservation policy. The first is a consolidation of relevant Congressional committees, and the second, a more adequate professional staffing of these committees.

Consolidation of Congressional Committees

Legislative proposals introduced in Congress are first referred to committees and considered by them; they may then be sent back to the chamber with recommendations for action. This system has grown greatly since the early days of the republic. Before the Legislative Reorganization Act of 1946, responsibility for legislative action was scattered among more than a hundred committees; even now (1951) there are more than fifty committees.[17]

Consolidation of Senate committees into thirteen has been proposed, of which eight would deal with substantive policy and five would have merely administrative functions.[18] The eight policy committees would be empowered to act jointly with corresponding committees of the House. Each of the policy committees would have twelve members and would be exclusive—that is, its mem-

---

of the Purse," *Political Science Quarterly*, June, September 1943, pp. 161–190, 380–414.

Robert Heller, *Strengthening the Congress* (Washington: National Planning Association, 1945), 41 pp. (Planning Pamphlets, No. 39).

[16] 60 *U. S. Stat. at L.* (1946), chap. 753.

[17] There are first the separate committees. At present there are nineteen in the House and fifteen in the Senate. Of the latter, six deal largely with natural resources (Agriculture and Forestry; Armed Services; Interior and Insular Affairs; Interstate and Foreign Commerce; Public Lands; and Public Works). Five deal with problems which have some more distant relation to resource policies (Appropriations; Banking and Currency; Expenditures in the Executive Departments; Finance; and Foreign Relations).

Second, there are the joint standing committees—for example, on atomic energy, on the economic report, on taxation, on printing, and on the library.

Third, from time to time both houses set up special or *ad hoc* committees for the consideration of particular problems. Problems of crime and of un-American activities are well-known recent examples. At present there are about a dozen of such special committees.

Fourth, there are many Congressional commissions and boards on various problems.

[18] S. Res. 169, 78th Cong., 1st sess.

bers would have no other major committee assignment.[19] The eight committees proposed are Agriculture and Forestry, Natural Resources and Public Works, Interstate Commerce, Finance and Monetary Affairs, Foreign Relations, Armed Forces, Labor and Public Welfare, and Judiciary. This proposal was only in part incorporated into the Act of 1946 mentioned above.

In order to avoid duplication and conflicts, the Agriculture and Forestry Committee might well deal with all flow resources, including water, grazing, fisheries, and recreation. The Natural Resources Committee might deal with all stock resources. These and the other policy committees would then parallel the individual sections of the planning and review division suggested for the executive branch of governments and the consolidation of responsibility suggested for executive departments.

Professional Staffing of Congressional Committees

Apart from the small professional staff associated with the Appropriations and Internal Revenue committees, the Office of the Legislative Counsel,[20] and the experts temporarily employed for special committee inquiries, the only specialists available to Congress within the legislative establishment are those supplied by the Legislative Reference Service (L.R.S.).[21] In spite of recent additions,[22] especially of economists, a further expansion of the L.R.S. would seem well worth the cost.

The question arises whether the policy committees should appoint their own staffs, as the Appropriations and Internal Revenue committees do, or whether specialists should be drawn from a permanent pool like the L.R.S. Although something may be said on both sides, the latter alternative appears preferable.

Scientific impartiality, competence, and nonpartisanship ap-

---

19 Many senators serve on so many committees that effective participation and even regular attendance are impossible. The situation in the House is somewhat better because of the greater number of members.

20 Assistance by this office is confined to bill drafting.

21 The Legislative Reference Service was created by Congress in 1919. It is attached to the Library of Congress.

22 Since the Legislative Reorganization Act of 1946, 18 senior specialists have been added to the staff of the L.R.S. Other classifications, such as legal analyst, special analyst, air-transport consultant, and research counsel, have also been added.

pear more effectively safeguarded if experts are appointed by the director of the L.R.S. in accordance with the requirements of the civil-service system. In consultation with committee chairmen, members of the L.R.S. can be assigned to policy committees in a way that insures the advantages of specialization in subject matter and continuity of service. The costs of an adequate staff for the L.R.S. are probably less than that of permanent separate staffs appointed by each policy committee.

Duplication Objectionable?

Some may protest that more intensive policy planning by legislative committees and their more adequate professional staffing will duplicate the work on policy planning which, in any event, must be done by the executive branch. Cases of such duplication may occur. In a system of checks and balances such as that of the government of the United States, some duplication in research and planning is unavoidable and, within limits, unobjectionable.

It is not clear whether and to what extent a shift of the locus of sovereignty from the legislative to the executive has taken place in recent decades, as is widely believed.[23] However, such a shift may best be prevented by helping Congress to act more effectively and intelligently. A high-class civil-service staff, directly employed by Congress, may be at least in part an answer to this problem.

Coördination in State Legislative Branches

Similar improvements in the committee system and in the professional staff available to legislators may be made in the legislative branches of state governments. Several state legislatures already have established committees, councils, or advisory

---

[23] U. S. Congress, House, *Recommendations and Proposed Legislation to Improve the Organization of Congress,* 7th Intermediate Report of the Select Committee of the House to Investigate Executive Agencies, 78th Cong., 2d sess., H. Rep. 1912 (Washington: 1944), 10 pp.

U. S. Congress, House, Committee on Expenditures in the Executive Departments, *Reorganization Plans Nos. 1, 2 and 3 of 1946,* Hearings, 79th Cong., 2d sess., on H. Con. Res. 151, 154, 155 (Washington: 1946), 341 pp.

boards which can employ a permanent staff in order to study proposed legislation and assist in the drafting of bills. Although these bodies are intended to be general executive committees of the legislature, especially if the legislature is not in session, their professional staff may, in due course, become the nucleus for legislative research pools from which may be drawn specialists to assist individual functional committees, particularly on resource development.

Much has been written lately about the alleged infringement of state rights by the federal government in the realm of natural resources. Regardless of whether such infringement has taken place or not, state rights may be defended best if the states themselves shoulder and discharge effectively the duties with which these rights are constitutionally associated.

More active participation of the states in conservation policy requires not only coördination of executive and legislative activities in each state but also federal–state and interstate coördination.[24] What are the possibilities for improvement in these two directions?

Federal–State Coördination

Some coördination in resource policies between the federal government and the states is brought about in Washington by congressmen of both houses representing state interests. This process, however, is rather haphazard and is influenced by many political and personal factors.

In some legislation—in the Flood Control Act of 1944, and the River and Harbor Act of 1946—it is stipulated that individual projects planned by federal agencies must be submitted to the states for review, and that the latter's comments must accompany requests for appropriations by the Congress. This legislation represents a desirable improvement over the Flood Control Act

[24] When federal–state and interstate relations are considered, the term "coöperation" would seem constitutionally more appropriate than "coördination" if the means rather than results are emphasized. For this reason the term "coöperation" was used above. When the term "coördination" appears in the following pages, it refers, strictly speaking, to the results of coöperation between constitutionally sovereign governments.

of 1938. It restored to the states that limited participation in the execution of individual flood-control projects which was granted to them in the original (1936) Flood Control Act.

There does not exist at present effective institutional machinery by which the states could act as consultants for the formulation of basic federal legislation in the resource field *before* such legislation is enacted by Congress. Neither is there institutional machinery through which the states could be consulted in a more continuous fashion than is possible under the two acts just mentioned and with respect to resource problems other than flood control. The concern here is with policy formulation and implementation, not with research. In research, federal agencies are more inclined to recognize the states as equal partners.

Attempts to Provide Better Federal–State Coördination

Attempts have been made to supply this needed institutional machinery. In agriculture, the Northern Great Plains Agricultural Advisory Council and its twin brother for the Southern Great Plains may be mentioned. However, the objective is coördination of research rather than of policies; active state participation is confined to the land-grant colleges.

Another type of machinery is provided by the Inter-Agency committees for the Missouri and the Columbia river basins, already mentioned. Representatives of the states are invited to the meetings of these committees,[25] but there is little participation of the states either in the meetings or in the work of important subcommittees.

Another attempt at better federal–state coöperation deserves a more detailed consideration as a first step in the right direction: this is the Pacific Coast Board of Intergovernmental Relations established in 1945, comprising California, Oregon, and Washington.

[25] For the Missouri Basin Committee, two representatives are appointed jointly by the governors of all basin states. For the Columbia River Committee, originally four joint representatives were invited. Later this was changed to a representative from each state.

The Pacific Coast Board of Intergovernmental Relations

The membership of the Pacific Coast Board of Intergovernmental Relations consists of the field chiefs of federal agencies as selected by the Pacific Coast Federal Regional Council mentioned above, and, from each member state, comprises the governor, the chairman of the State Commission on Interstate Coöperation, a representative of the official organization of cities and municipalities, and a representative of the organization of county supervisors or commissioners. No alternates are permitted. Any member may bring with him to any meeting such consultants as he may desire, provided such consultants are officials of the constituent jurisdictions. Regular meetings are held quarterly, alternating between Washington, Oregon, and California. The governor of the state in which the meeting is held is the chairman of the meeting. The principles of organization contain the provision that the membership from the federal government "shall be less in number than the combined total of representatives of the other jurisdictions." [26]

Although the purposes of the board are only discussion, and any conclusions are purely advisory, it has aided in the elimination of conflicts in the execution of existing local, state, and federal laws and regulations and has helped in the pooling of facts and exchange of ideas for the planning of government action concerning some vital economic and social problems. So far, most of the attention of the board has been given to the problems of reconversion and postwar adjustment, which were of special concern for the Pacific Coast. But there is no reason why problems of resource conservation could not be attacked through this institutional machinery or through one which may, in due course, be developed from it.

---

[26] The Pacific Coast Board of Intergovernmental Relations, *Principles of Organization* (San Francisco: The Pacific Coast Board of Intergovernmental Relations, 1945), p. 1.

Regional Organization of Federal Activities

It is no accident that attempts to create institutional mechanisms for better federal–state coöperation such as those just mentioned have started on a regional rather than a national basis. Most government responsibilities not vested entirely in the federal government by the Constitution are concerned with economic problems which are largely regional or have important regional characteristics. This is especially true for problems in resource conservation. The federal government could, therefore, facilitate federal–state coöperation in the resource field by adopting, to a greater extent than heretofore, a regional organization with more authority delegated to regional executives.

It is important for federal–state coöperation that the federal regions be chosen appropriately. At present, these federal regions comprise commonly four to six states, and differ considerably among different federal agencies. The need for coördination in conservation policies, phases of which are implemented by different federal departments, requires identical regions for such departments and identical headquarters. It is less clear what the proper size of federal regions should be.

Size of Federal Regions

There are several arguments in favor of the present rather large regions. Problems of resource conservation are usually not confined to one state. Membership in a common region will help individual states to think about their problems in regional terms and to coöperate. Participation of several states in the solution of a common problem will reduce the danger of a dominant influence by local political pressure groups. From the standpoint of central federal administration, resource policies can be more effectively and cheaply coördinated if regions do not become too numerous.

On the other hand, there are several arguments in favor of making the federal regions smaller. These arguments even suggest that for some states, the federal regions may eventually become identical with the states. Obviously, this should not be considered for small states with very similar problems—for example, the New England states. What are some of these arguments?

Growing responsibility of government on the federal as well as on the state level will probably call for greater intensity (besides better coördination) in federal–state relations. Identity of federal and state administration with respect to boundaries and seat of government would facilitate close contact. In some cases another point is important: individual regional problems in resources, and in economic development generally, may extend over different geographical areas. A small basic unit (a state) of federal regions would facilitate coöperation between different combinations of regions as required by the geographical extent of different problems. An example taken from empirical evidence may illustrate this point.

An Illustration

In some problems (transportation, migration, national defense, foreign trade, fisheries) the states bordering the Pacific form a logical unit. The Pacific Coast Board of Intergovernmental Relations, mentioned above, grew out of these relations. It had a predecessor in the Ninth Regional Civilian Defense Board as a medium through which the various levels of government discussed and solved many war-related civilian activities.

In other problems (agriculture, water development, mineral resources, forestry) the Pacific Northwest, including states without direct access to the Pacific (Montana, Idaho, Wyoming) and excluding California, constitutes a better grouping. The Columbia Basin Inter-Agency Committee and the Northwest States Development Association are outgrowths of these interrelations.

In still other respects (livestock economy, financial connections, domestic trade, tourist business) California is closely related to its hinterland in Nevada, Arizona, and Utah. These states in turn are tied together with Colorado and New Mexico through the Colorado River and similarity of climate, topography, and history. Such interrelations have led to the Colorado River compact.

Nonconstancy of Regional Relations

These interrelations do not necessarily remain constant. Physiographic factors are, of course, largely constant; but their relative economic weight varies with changes in methods of communication (railroads, roads, automobiles, airplanes, radios) and of water development (navigation, irrigation, power and flood control, growth of a unified system in distributing electricity).

Flexibility in the geographical configuration of a region may, therefore, be desirable. Smaller permanent basic units of federal administration would facilitate such a flexibility.

Conclusions Concerning the Size of Federal Regions

It is rather difficult to draw conclusions from these pros and cons of whether federal administrative regions should be large or small. A region consisting of a small number of states, let us say two or three, or at most four, would seem at present to be the best compromise between the various arguments set forth above. Ultimately the more important states may each become a federal region.

Interstate Coördination

Turning now to the problem of better interstate coördination in resource policies, we may begin here also by noting some past attempts to set up suitable institutional machinery. In this case, as in that of federal–state relations, the term "coördination" refers to the results of voluntary coöperation between constitutionally sovereign governments.

The Council of State Governments has been active here. Among its achievements one may mention the organization of the Atlantic States Marine Fisheries Commission, the Interstate Commissions on the Delaware and the Potomac basins, the Arkansas-Oklahoma Interstate Water Resources Committee, the Missouri Valley States Committee, the Northwest States Development Association, and numerous other agencies created by state compacts. Although most of the agencies created (with or without state compacts) have only investigating and advising functions, the experience with this type of interstate coöperation has been rather encouraging.

In the future, the Council of State Governments may well consider whether facilitating interstate coöperation in the planning and execution of conservation policy should not constitute one of its major objectives. Among state functions, regulation of resource use looms large and is likely to become even more important in the future. Depletion of resources, rapid technological development, and increasing and shifting demand will create rather complex and pressing conservation problems. Regulation of resource use is constitutionally left to the states. Historically, the states have been closely connected with the conservation movement since the White House Conference of Governors in 1908. Many resource problems concern more than one state, and these problems can be attacked most effectively through coöperative action by all states concerned.

Through the commissions on Interstate Coöperation, the Council of State Governments already possesses a mechanism through which efforts toward better interstate coöperation in the resource field could be channeled. The council was instrumental in holding conservation conferences among the eastern states. Special committees to study resource problems—for example, in forestry, in agriculture, in water resources, in postwar reconstruction and development—have been set up. The council has aided in the establishment of state compacts. Some of the permanent interstate agencies created by these compacts were mentioned.

Activity of the council may well be intensified in these three directions, namely, in sponsoring regional and national conservation conferences, in setting up regional and national study committees, and in establishing permanent interstate and federal–state advising or administrative agencies concerned with specific regional resource problems. To aid in these activities the existing, rather small professional staff of the Council of State Governments would need expansion. A permanent interstate resources committee may be helpful in guiding the council's activities in this field and in representing the views of the states before the public and before the federal government.

Among the many specific resource problems (besides the general problems of better interstate and federal–state coördination in resource policies) for study and recommendation by the council, the following may be mentioned especially: (1) taxation, credit, and tenure of resources (including resources in public ownership); (2) integrated, multiple-use development of water resources; (3) state forestry legislation; (4) conservation of wildlife (fisheries, upland game, migratory birds); (5) regulation of mining and drilling operations; (6) administration of grants-in-aid, especially in soil conservation; and (7) the function of special districts (for example, soil conservation, forestry, grazing, irrigation, fire protection) in resource conservation.

State Compacts

Although state compacts are not a necessary condition for better interstate coöperation in conservation policies, they are generally helpful. The experience gained in negotiating, concluding, and administering state compacts has resulted in a better integration of state policies, even for resources which were not directly affected. This educational by-product is not the least contribution of state compacts for interstate coöperation. In the past, use of state compacts has been confined largely to water, oil, and natural gas. In many cases, conservation of land, forests, grazing, minerals, and wildlife can be aided by extending the use of state compacts to these resources.

Unified River Basin Authorities

Our survey would not be complete without mentioning the current, somewhat controversial, attempt to solve the problems of coördinating resource policies by one stroke: by setting up through federal legislation unified planning-administrative-construction agencies for the major river basins of the country modeled after the Tennessee Valley Authority.

In view of existing weakness in intrafederal, federal–state, and interstate coöperation, the TVA constitutes a valuable experiment. Moreover, as already mentioned, the existence of the TVA and proposals for other basin authorities have prompted some federal agencies to coöperate with each other and with the states. This "yardstick effect" of the TVA upon federal executive agencies is at least as important as its more frequently discussed yardstick effect upon privately owned public-utility companies.

Whether the TVA is a model that should be used in other river basins depends on whether there are no better alternatives by which coördination and coöperation in resource policy could be improved in the future. At best, independent administrative authorities for major watersheds would solve the problem of policy coördination only for individual geographic areas and only for a limited sphere of conservation policy. In doing so they would take over some of the functions of regular legislative and executive machinery (federal, state, local), which cannot be dispensed with in any event. The essential problems involved in the coördination of conservation policy in a unified way for regions, and for the country as a whole, would not be solved and might be complicated and delayed. On the other hand, if these problems could be solved, separate authorities for each river valley would be unnecessary.

Furthermore, as has been indicated above, regions, from the standpoint of conservation policy, are not necessarily identical with watersheds, and the economic significance of natural features may change. Technological development has generally tended to decrease this significance.

Railroads, hard-surfaced roads, automobiles, airplanes, and radios have decreased the relative importance of watersheds for economic differentiation.

More and more, technology of water-resources development itself tends in the same direction. Navigation has decreased in relative importance. Water for power, irrigation, and municipal uses has become relatively more important. Water power is transported in the form of electricity from one watershed to the other. Increasingly, water for irrigation and municipal use is so transported—for example, from the Colorado River to the Pacific Coast, from the Sacramento Valley into the San Joaquin Valley, and from watersheds west of the Continental Divide to those east of it (Gunnison-Arkansas project). Even greater projects of water transportation are under consideration—for example, from the Columbia River watershed to California, from the Trinity and Klamath rivers in northwestern California to the Central Valley of California and, indirectly, that is, through water exchange, to the south coast of California.

We may conclude that modern technology tends to render a natural watershed obsolete as an independent unit for planning, administration, and construction. Improvement in coördination and coöperation between regular executive and legislative agencies at various governmental levels appears, in the long run, more essential for implementation of conservation policy than the creation of special planning-administrative-construction authorities organized on a watershed basis.

Appendix

MATHEMATICAL NOTE ON THE THEORY OF CONSERVATION

1. Definition of Conservation

If $x_1, x_2, x_3, \ldots x_n$ are rates of use of the resource X in planning intervals (see definition in chapter 3) $t_1, t_2, t_3, \ldots t_n$, we have conservation, neutrality, or depletion according to whether $\Delta x_1 + 2\Delta x_2 + 3\Delta x_3 + \ldots + n\Delta x_n$ is greater than, equal to, or less than zero. Why the first power of time is used for weighting is discussed in chapter 4.

The degree of conservation (depletion) is defined:

$$\frac{\Delta x_1 + 2\Delta x_2 + 3\Delta x_3 + \ldots + n\Delta x_n}{x_1 + 2x_2 + 3x_3 + \ldots + nx_n}$$

The average weighted change of rates of use is defined:

$$\frac{\Delta x_1 + 2\Delta x_2 + 3\Delta x_3 + \ldots + n\Delta x_n}{1 + 2 + 3 + \ldots + n}$$

The average weighted distance of changes in rates of use is defined:

$$\frac{\Delta x_1 + 2\Delta x_2 + 3\Delta x_3 + \ldots + n\Delta x_n}{\Delta x_1 + \Delta x_2 + \Delta x_3 + \ldots + \Delta x_n}$$

The average weighted rate of use is defined:

$$\frac{x_1 + 2x_2 + 3x_3 + \ldots + nx_n}{1 + 2 + 3 + \ldots + n}$$

The average weighted distance of rates of use is defined:

$$\frac{x_1 + 2x_2 + 3x_3 + \ldots + nx_n}{x_1 + x_2 + x_3 + \ldots + x_n}$$

The concepts of average weighted distance of rates of use and average weighted distance of changes in rates of use play an important role in interest theory. They may also be used for defining conservation and depletion. Such a definition, however, would not be useful when considering some important special cases, for instance, an increase or a decrease of a constant rate or a rate with constant fluctuations—problems that occur particularly in agriculture, forestry, and grazing.

2. The Optimum State of Conservation

If V stands for present net revenues, R for present revenues, and Q for present costs, we have:

$$V = R(x_1, x_2, x_3, \ldots x_n) - Q(x_1, x_2, x_3, \ldots x_n)$$

Differentiating, we obtain as necessary condition (for sufficiency, well-known auxiliary requirements must be met) for the optimum state of conservation:

$$\frac{\delta}{\delta x_1} R(x_1, x_2, x_3, \ldots x_n) = \frac{\delta}{\delta x_1} Q(x_1, x_2, x_3, \ldots x_n)$$

$$\frac{\delta}{\delta x_2} R(x_1, x_2, x_3, \ldots x_n) = \frac{\delta}{\delta x_2} Q(x_1, x_2, x_3, \ldots x_n)$$

$$\cdot \quad \cdot \quad \cdot \quad \cdot \quad \cdot \quad \cdot \quad \cdot \quad \cdot \quad \cdot \quad \cdot$$

$$\cdot \quad \cdot \quad \cdot \quad \cdot \quad \cdot \quad \cdot \quad \cdot \quad \cdot \quad \cdot \quad \cdot$$

$$\frac{\delta}{\delta x_n} R(x_1, x_2, x_3, \ldots x_n) = \frac{\delta}{\delta x_n} Q(x_1, x_2, x_3, \ldots x_n)$$

If $v_1, v_2, \ldots, v_n$ and $r_1, r_2, \ldots r_n$ and $q_1, q_2, \ldots q_n$ denote future net revenues, revenues, and costs in intervals $t_1, t_2, \ldots t_n$, and i the interest rate, we may write :

$$v_1(x_1, x_2, \ldots x_n) = r_1(x_1, x_2, \ldots x_n) - q_1(x_1, x_2, \ldots x_n)$$

$$v_2(x_1, x_2, \ldots x_n) = r_2(x_1, x_2, \ldots x_n) - q_2(x_1, x_2, \ldots x_n)$$

$$\cdot \quad \cdot \quad \cdot \quad \cdot \quad \cdot \quad \cdot \quad \cdot \quad \cdot \quad \cdot \quad \cdot \quad \cdot$$

$$\cdot \quad \cdot \quad \cdot \quad \cdot \quad \cdot \quad \cdot \quad \cdot \quad \cdot \quad \cdot \quad \cdot \quad \cdot$$

$$v_n(x_1, x_2, \ldots x_n) = r_n(x_1, x_2, \ldots x_n) - q_n(x_1, x_2, \ldots x_n)$$

Present total functions in terms of future total functions may be written:

$$V = v_1(1+i)^{-1} + v_2(1+i)^{-2} + \ldots + v_n(1+i)^{-n}$$

$$R = r_1(1+i)^{-1} + r_2(1+i)^{-2} + \ldots + r_n(1+i)^{-n}$$

$$Q = q_1(1+i)^{-1} + q_2(1+i)^{-2} + \ldots + q_n(1+i)^{-n}$$

Present marginal functions in terms of future marginal functions, for example, with respect to x_1 may be written:

$$\frac{\delta}{\delta x_1} V = \frac{\delta}{\delta x_1} v_1(1+i)^{-1} + \frac{\delta}{\delta x_1} v_2(1+i)^{-2} + \ldots + \frac{\delta}{\delta x_1} v_n(1+i)^{-n}$$

$$\frac{\delta}{\delta x_1} R = \frac{\delta}{\delta x_1} r_1(1+i)^{-1} + \frac{\delta}{\delta x_1} r_2(1+i)^{-2} + \ldots + \frac{\delta}{\delta x_1} r_n(1+i)^{-n}$$

$$\frac{\delta}{\delta x_1} Q = \frac{\delta}{\delta x_1} q_1(1+i)^{-1} + \frac{\delta}{\delta x_1} q_2(1+i)^{-2} + \ldots + \frac{\delta}{\delta x_1} q_n(1+i)^{-n}$$

Necessary conditions (here again, for sufficiency well-known auxiliary requirements must be met) for the optimum state of conservation in terms of present as well as future net revenues are:

$$\frac{\delta}{\delta x_1}V = \frac{\delta}{\delta x_1}v_1(1+i)^{-1} + \frac{\delta}{\delta x_1}v_2(1+i)^{-2} + \ldots + \frac{\delta}{\delta x_1}v_n(1+i)^{-n} = 0$$

$$\frac{\delta}{\delta x_2}V = \frac{\delta}{\delta x_2}v_1(1+i)^{-1} + \frac{\delta}{\delta x_2}v_2(1+i)^{-2} + \ldots + \frac{\delta}{\delta x_2}v_n(1+i)^{-n} = 0$$

$$\cdot \quad \cdot \quad \cdot \quad \cdot \quad \cdot \quad \cdot \quad \cdot \quad \cdot \quad \cdot \quad \cdot \quad \cdot \quad \cdot$$

$$\cdot \quad \cdot \quad \cdot \quad \cdot \quad \cdot \quad \cdot \quad \cdot \quad \cdot \quad \cdot \quad \cdot \quad \cdot \quad \cdot$$

$$\frac{\delta}{\delta x_n}V = \frac{\delta}{\delta x_n}v_1(1+i)^{-1} + \frac{\delta}{\delta x_n}v_2(1+i)^{-2} + \ldots + \frac{\delta}{\delta x_n}v_n(1+i)^{-n} = 0$$

3. Relations between Use Rates and between Resources

As before, $V(x_1, x_2, x_3, \ldots x_n)$, $R(x_1, x_2, x_3, \ldots x_n)$ and $Q(x_1, x_2, x_3, \ldots x_n)$ are the total present net revenue, revenue, and cost functions of rates of use in planning intervals $t_1, t_2, t_3, \ldots t_n$.

Two rates, for instance x_1 and x_2, are defined as complementary, competitive, or independent in net revenues according to whether $\frac{\delta^2 V}{\delta x_1 \delta x_2}$ is greater than, less than, or equal to zero.

The same definitions are used in the case of revenues.

In the case of costs, two rates are defined as complementary, competitive, or independent according to whether $\frac{\delta^2 Q}{\delta x_1 \delta x_2}$ is less than, greater than, or equal to zero.

The second cross partial derivatives are likewise used for defining complementarity, competitiveness, and independence of rates of use of different resources (in net revenues, revenues, and costs) in any planning interval which may be considered. This means that rates of use of other resources—for example, $y_1, y_2, y_3, \ldots y_n$, and $z_1, z_2, z_3, \ldots z_n$ of two resources Y and Z—are introduced as additional variables in net revenue, revenue, and cost functions.

INDEX

Accessible resources, 336–39
Ad-valorem taxes, 186
Adams, R. L., 153
Adjustment to economic fluctuations, 203–12
Administrative valuation, 239–44
Administrative weighting, 245
Advertising, 191
Aereboe, Friedrich, 182
Agricultural Adjustment Administration, 135, 279, 280, 300, 331, 356
Agricultural Extension Service, 278–80
Agriculture, adjustments to economic instability in, 203–9; commodity agreements in, 323, 326–31; conservation education in, 277; cost relations in, 72; flexible plans in, 118; imperfect markets in, 102; overexpansion in, 323, 330; planning intervals in, 80; safe minimum standard in, 258–59; taxes in, 175
Allocation, 147–49, 151, 153–55, 235–39, 283
American-Canadian Sockeye Salmon Convention, 310
American Political Science Association, 365
Amortization of initial costs, 79
Amortization payments, 145, 162
Anglo-American Oil Agreement, 332
Anticyclical conservation policy, 212–16
Animal species, 256, 258
Appraisal, obsolete practices in, 166
Apprenticeships, 281, 344
Approximations, 21, 24, 78, 80–83, 249–50
Arkansas Oil and Gas Conservation Law, 297
Army Engineers, *see* U.S. Corps of Engineers
Assessment, inequalities of, 183; deferred reassessment, 184
Assets, forms of, 100; imperfect markets for, 101–2, 157, 205; value of, 178
Assumptions, 21, 130, 169, 174, 190
Atomic energy, how to classify, 42
Atomic Energy Commission, 301, 364

Bacher, Robert F., 42
Bacon, L. B., 329
Bacon, Robert, 314, 315
Balance of payments, international, 216
Bankhead-Jones Tenant Farms Purchase Act, 159, 163, 217
Barber, E. Lloyd, 172
Barna, Tibor, 185
Bartz, Patricia McBride, 144
Basset, Edward M., 284

Bauxite, 37, 209, 210, 331
Baxter, Charles H., 178
Beneficiation processes, 6
Beard, Charles A., 227
"Best knowledge," 56
Bilateral monopoly, 190, 191
Birds, 142, 217, 306, 310–11, 313
Birth control, 8, 16, 19, 341–42
Birth rates, changes in, 8
Blighted areas in cities, 286
Bonneville Power Administration, 359
Bovine tuberculosis eradication, 285, 301
Bowman, Robert G., 330
British Association for the Advancement of Science, 320–21
Buffer stocks, 218, 327, 329, 332, 339
Business cycles, 200, 202–19

Calculus of variations, 78, 233
California, conservation education in, 278; forest districts in, 293–96; leases in, 152, 154; oil pools in, 143; regulation of water use in, 287–90
California Civil Code, 288
California Community Redevelopment Act, 286
California Conservation Council, 278
California Constitution, 288
California Division of Water Resources, 288–90, 364
California Forest Conservancy Law, 293
California Forest Practices Act, 294
California Oil and Gas Conservation Act, 143, 297
California State Board of Forestry, 294, 296
California Water Code, 288, 289
Capital formation, in underdeveloped countries, 345
Capital-gains tax, 172, 175
Capitalism, and uncertainty, 124; *see also* Industrial societies
Capitalistic revolution, in social institutions, 13
Capitalization of taxes, 174
Cartels, 198, 308, 309, 322, 331; *see also* Commodity agreements; Monopolistic conditions
Central Valley project, 240
Chamberlin, Edward, 192
Charges, fixed, progressive, and regressive, 106; flexible, 108
Civilian Conservation Corps, 215, 280–81
Classification, of resources, 29, 33, 35, 36, 38, 40, 42, 43
Coal, 36, 176, 264–65, 298, 331
Collateral, obsolete practices with, 167
Collective extramarket goods, 240–44
Collective revenues and costs, 236
Colonial exploitation, 9, 200, 341–42; *see also* Plantation agriculture
Columbia Basin Inter-Agency Committee, 370, 373
Committee on Price Determination, 205
Commodity agreements, 218, 305, 321–34
Commons, John R., 226
Compensation, in leases, 152, 154; with regulations, 300–1; with zoning, 285, 286
Competition, monopolistic and pure, 69, 189–98, 234
Competitiveness and complementarity, between resources, 44, 262, 381; between use rates, 66–75, 133, 171, 194, 381
Congress (U.S.), organizational reforms, 365–68
Congressional committees, 366–68
Connally Oil Transportation Act, 297
Conference of Governors on Conservation of Natural Resources, 314
Conferences, international, 305–20, 325
Congress of Agriculture and Fisheries, 307
Conservation, accomplishing, 56; adoption of the term, 27; definition, 51, 53, 379–80; dynamic interpretation, 51; economic forces affecting, 95–219; economic limits of, 58; *ex ante* and *ex post*, 54, 99, 170; meanings, 19, 28, 48–52; measurement, 52, 55; modern problem of, 15; state of, 53; *vs.* investment, 58; wasteful, 57
Conservation decisions, 66–75
Conservation movement, beginnings, 15, 314
Conservation policy, and other policies, 24, 348–49; and social institutions, 223–29; anti-cyclical, 212–16; confused with other objectives, 20, 316; coördination of, 347–78; criteria, 230–50; definition, 223; domestic, 216, 271–303; formulation and execution, 224, implementation, 271–346; international, 217–19, 304–46; objectives, 230–50, 251–67, 337, 341; on tenancy,

151–58; opportunities for, 109; to compensate for unequal income distribution, 246; tools, *see* Tools; urgent for flow resources, 47; with critical-zone resources, 251–67
Conservation practices, concept, 82; to avoid the critical zone, 258–59
Conservation theory, mathematics of, 379–81; *see also* Maximization theory
"Conserving" products, 133–35, 169, 280
"Conserving" services, 132, 169
Consolidation of conservation agencies, 347–51, 359–65
Constitutional framework of policy, 351–53
Consumer expenditures, 246–47
Consumer sovereignty, 248
Contraction of production, 203–12, 216, 218, 323
Controls, *see* Direct tools; Penalties; Regulation; Subsidies; Zoning
Conventions and treaties, 5, 217, 305–20, 325, 326
Cook, W. W., 150
Coombs, Whitney, 183
Coördination of governmental agencies, 24, 347–78
Correlative Rights Doctrine, 144, 287
Corry, Andre V., 210
Cost structure, in agriculture, 203–9; in mining, 209–12
Costs, and exhaustibility, 33; factors affecting changes of, 34; in income tax, 175; incidence *vs.* allocation of, 147–49, 151, 153–55; redistributing, 98; private and social, 231–38; relations through, 67, 70–75, 133, 381; role of, in adjustments to economic fluctuations, 203–12; *see also* Lumpy costs; Marginal costs; Recovery costs; Sunk costs
Council of State Governments, 374–75
County regulations, 287, 290, 292, 298
Court decisions, on regulations, 299; on water use, 288–90; on zoning, 284, 285
Courts, for debt adjustment, 162; of tenancy arbitration, 154; tenure, 162
Credit, 161–67; and property rights, 141, 161; imperfect markets in, 101, 166, 195, 202, 205; *see also* Interest rates
Credit agreements, conservation clauses in, 167
Credit insurance, 166
Credit risks, 163, 166
Creditor's rights, 161
Critical zone, 38–39, 256–57; already passed, 265–66; and habit patterns, 127
Critical-zone resources, 39–43, 252, 253; competitive relations, 263–65
Crop insurance, 121–24
Crops, "conserving" and "depleting," 133–35
Cumulative use, concept and measurement, 31; effects of, on revenues and costs, 34; maximum, 60
Cushman, Robert, 353
Customs, leasing, 152, 154; market, 188
Cutting practices, 295
Cycles, business, 200, 202–19; production, 201

Death rates, changes in, 8, 11, 341–42
Death taxes, 184–85
Definitions, 19, 27, 28, 29, 31, 32, 33, 35, 37; mathematical, 379–81; *see also* specific words
Demand, complementarity and competitiveness in, 44, 262–64
Demand curves, and monopoly, 191; during depressions, 211
Demand functions, 231, 241
Demand schedules, 241, 242
Democracy, property rights in a, 148; social weighting in a, 245; valuation in a, 244
"Depleting" products, 133–35, 169
"Depleting" services, 132, 169
Depletion, and economic instability, 199–212; as an objective, 136, 177, 181; definition, 51, 53, 379; early diagnosis of, 275–77; explanations of, 15; from natural causes, 266–67; in colonial exploitation, 9, 200; in industrial societies, 3–14; irreversible, 39, 251–53; not always wasteful, 57; of domestic strategic resources, 337; *vs.* disinvestment, 58
Depletion allowances, in income tax, 176
Depressions, depletion during, 104, 106, 108, 203–12; effects of, on time preferences, 104, 202; fixed charges during, 108
Diagnosis of depletion, 275–77
Dickinson, H. D., 239
Dictatorships and uncertainty, 125
Direct tools of conservation policy, 23, 229, 272–303

Discontinuous changes, 80–81
Discounting, for taxes, 180; for time, 77, 98, 104, 179; for uncertainty, 87, 113, 115, 126
Discovery of new stocks of resources, 323
Disinvestment *vs.* depletion, 58
Dispersion, *see* Probability distribution
Distance of use rates, definition, 379
Distribution, time, *see* Time distribution
Districts, special, 287, 290, 291–96
Diversification, 124, 208
Dobzhansky, Theodosius G., 252
Domestic tools of conservation policy, 271–303; *see also* Tools
Dreesen, W. H., 183
Drought insurance, 121
Dupuit, Jules, 232
Dyopoly, 69, 190, 193

Economic analysis, as an objective, 20
Economic calculations, *vs.* habit patterns, 92
Economic fluctuations, 199–219
Economic forces, 95–219; and social institutions, 139; as tools of policy, 109; why analyzed, 97; *see also* Credit; Economic instability; Income; Interest; Market form; Penalties; Price supports; Prices; Property rights; Social institutions; Subsidies; Taxation; Tenancy; Uncertainty
Economic instability, 199–219, 323–24; meaning, 201; mitigating effects of, 216; *see also* Price instability
Economic institutions, 141; *see also* Social institutions
Economic theory, *see* Conservation theory; Instantaneous economics; Maximization theory; Time Economics
Economics of conservation, aims, 18
Edminster, Lynn R., 331
Education, 274–82, 302; not a cure-all, 281; in underdeveloped countries, 343–45
Educational agencies, 278–81
Efficiency, *vs.* conservation, 57
Elasticity of price expectations, 131
Elasticity of supply, 231; in agriculture, 203, 206; in mining, 210
Ely, Northcutt, 144
Emergency Relief Act, 217
Eminent domain, 284, 285, 286, 299
Ends-means scheme, 28, 29, 37
Energy, constancy of, 16
Englund, Eric, 183
Entrepreneurial revenues and costs, 236–38
Equalitarian ideology, 245
Equations, 379–81
Erosion, geologic, 266–67
Estate taxes, 184–85
Europe, soil fertility in, 10; tenancy in, 152, 153, 154
European Timber Exporters' Convention, 325
Ex ante and *ex post* appraisal of habit patterns, 91
Ex ante and *ex post* concepts, of conservation, 54, 99, 170; of optimum state, 77, 83, 89, 91, 93; of resources, 30, 32
Ex ante and *ex post* effects, of depressions, 203–12; of interest, 99
Excess profits tax, 177
Executive branch of government, 224, 351–65, 369–76
Executive Office of the President, 354
Exhaustibility of resources, 33, 34
Exhaustion of resources, *see* Depletion
Expansion periods, 200, 202–19
Expectations, of price changes, 131; of tax changes, 170; probability distribution of, 86, 111
Extramarket goods, valuation, 85, 239–48

Fair Trade Practices Act, 188
Fairchild, Fred Rogers, 180, 183
Farm Credit Administration, 166
Farm Security Administration, 165, 166
Farm tenancy, *see* Tenancy
Farmers Home Administration Act, 158, 164
Federal conservation agencies, 349, 352–58, 359–65, 372–74, 376–78
Federal Crop Insurance Act, 121
Federal forests, regulations of, 293
Federal Inter-Agency River Basin Committee, 357, 370
Federal Land Bank of St. Louis, 167
Federal government, as landlord, 158
Federal Oil Conservation Board, 143
Federal Power Commission, 353, 357
Federal regions, 372–74
Federal-state coördination, 369–74
Federal Trade Commission, 188, 198, 353

Feis, Herbert, 331
Fellowships, for training, 281, 344
Fernow, Bernhard E., 33
Finch, J. W., 348
Fire districts, 293
Fire-prevention practices, 295
Fish, and hydraulic mining, 298; international conservation of, 306–10, 313
Fisheries, 60, 142, 145, 217, 306–10, 313
Fishing, regulation of, 296, 361, 365
Fixed charges, 106, 107, 155, 162, 209
Fixed costs, 71, 204, 209, 210
Fixity, 106, 107; of rents, 155
Flexible charges, 108, 156, 163–64
Flexible plans, 87, 113, 116–19, 126
Flood control, 267
Flood Control Act, 235, 357, 369
Flood-control districts, 292
Flow resources, classes of, 38–43; commodity agreements in, 324–31; consolidating the agencies for, 350–51, 353, 359–65; defined, 37; depletion of, 3, 4; international tools for, 305, 306–20, 334; market forms in, 69; social institutions important for, 46; storage of, 37; taxes on, 172; with and without a critical zone, 38–43, 46
Fluctuations, economic, 200, 202–19
Food and Agriculture Organization, 319–20
Forces, economic, *see* Economic forces
Foreclosure, 161
Foreign policy, and conservation, 334–46
Foreign resource development, 337, 340–42
Forest credit system, 166, 167
Forest districts, 294–96
Forest resources, critical zone in, 257; regulations for, 293–96; safe minimum standard in, 258, 259; zoning for, 282
Forestry, cost relations in, 73; planning intervals in, 80; sudden liquidations in, 323, 326; taxes in, 172, 175, 185, 186; technological changes in, 5; *see also* Timber
4-H Clubs, 278
Freezing of strategic resources, 337, 338
Frisch, Ragnar, 233
Fugitive resources, 75, 141–45, 196, 217; international, 306–14
Functions, 380–81
Fungibility, imperfect, 101, 102
Fur Seal Convention, 308
Future net revenues, discounting, 98; equation for, 380; taxes on, 178

Gabbard, L. P., 183
Gamblers, uncertainty preference of, 112
Game, depletion of, 5; *see also* Wildlife
Gas (natural), 142, 143, 297; taxes on, 176
General property tax, 177–84
Geneva Convention, 308
Geneva Economic Conference, 325
Goodfellow, D. M., 90
Goods, durable, as a Flow resource, 41
Government, as creditor, 167; as landlord, 158; branches of, 224, 351–53
Governmental agencies, coördination of, 24, 347–78
Governors, White House Conference of, 314, 375
Gray, L. C., 33, 154
Grazing, time and degree, 257
Grazing districts, 296; *see also* Taylor Grazing Act
Grazing lands, as fugitive resource, 144; public, 158; regulation of use of, 296; safe minimum standards in, 258, 259; *see also* Range
Gross-revenue taxes, 186
Grotius, Hugo, 306
Ground water, a fugitive resource, 142, 144; critical zone in, 256; depletion, 5; regulations of use of, 288; safe minimum standard in, 258, 259; zoning for, 282

Habit patterns, and a safe minimum standard, 260; and birth rates, 342; and concept of optimum state, 89; and social institutions, 226; and uncertainty, 126; effect on conservation, 127; *ex post* appraisal, 91; in modern societies, 90; in underdeveloped societies, 109; *vs.* economic calculation, 92
Hahn, Eduard, 227
Halibut, 309
Halibut Production Control Board, 309
Hamilton, Walton H., 226
Hammar, C. N., 348
Harding, S. T., 290
Hardy, C. O., 200, 321
Havemeyer, Loomis, 314

Hayden, Sherman S., 311, 312
Hendrickson, C. I., 300
Hardwicke, R. E., 144
Hedging, 113, 119
Helium, 299, 301, 335
Heller, Robert, 366
Henderson, Alexander, 233
Hertzler, Joyce O., 226, 227
Hess, Ralph H., 33
Hicks, J. R., 233
Higgins, Benjamin, 213
High-seas fisheries, *see* Fisheries
Holmes, Oliver Wendell, 285, 299
Homo economicus, 84
Hoover Commission, 347, 353, 357, 358, 361–63
Hotelling, Harold, 196, 232, 233
House of Representatives, committees, 366–68
Hunting, regulation of, 296
Hydraulic mining, 264–65, 298
Hydroelectric power, 350, 359–60, 363

Imbalance of property rights, 147–49
Imperfect knowledge, 275
Imperfect markets, 101–3, 155, 157, 166, 189, 195, 202, 205
Implementation of policy, *see* Tools
In-training, 281, 344
Inaccessible resources, 336, 339–40
Incidence, 147–49, 151, 153–55, 235–39, 241, 243
Income, as an economic force, 104–9
Income distribution, effect on conservation, 246–48
Income level, effect on conservation, 105; effect on time preferences, 104, 202
Income supports *vs.* price supports, 107, 137
Income taxes, 172, 174–77
Indebtedness, 162
Indefiniteness of property rights, 141–45
Independent resources, 44
Independent use rates, 68, 381
Indifference curves, revenue and cost, 77
Indirect tools of conservation policy, 272; *see also* Economic forces; Social institutions
Industrial societies, depletion in, 3–14; *see also* Capitalism
Industrialization of underdeveloped countries, 343–46
Industry, meaning, 190
Inheritance taxes, 184–85
Inputs, changing time distribution of, 56
Insect and pest-control regulations, 293
Instability, economic, 199–219, 323–24; of government, 124, 145, 146; of property rights, 145–46, 152, 162
Instant of time, concept of, 31
Instantaneous economics, 71, 105, 190–93, 204–6, 232–33
Instantaneous rate of flow, mathematical definition, 31
Institute of Inter-American Affairs, 318
Institutional lags, 166
Institutions, *see* Economic institutions; Social institutions
Insurance, 115, 120–24
Insurance premiums, 116; subsidized, 122–24
Intangible values, 85
Inter-American Conference on Conservation of Renewable Resources, 318
Interest, 97–109
Interest payments, fixed *vs.* flexible, 108, 163
Interest rates, effect on conservation, 98; in time discounting, 79, 98; in uncertainty discounting, 115–16; market, 100, 195, 202; negative, 98; regressive, 107
International agencies, 305, 307–12, 317–19; proposed, 319–20
International Bank, 167
International Bureau of Whaling Statistics, 308
International Commission for the Protection of Wildlife, 312
International commodity agreements, 218, 305, 321–34
International conferences, conventions, congresses, and treaties, 5, 217, 305–20, 325, 326
International Conservation Congress, 314
International conservation policy, 217, 337, 341, 347, 352
International Convention for the Protection of Fur Seals, 308
International Council for the Study of the Sea, 307
International Fisheries Commission, 309
International Fisheries Exhibition, 307
International Forestry Center, 319

International Institute of Agriculture, 311, 319
International Labour Code, 333
International Ornithological Congress, 310
International Pacific Salmon Fisheries Commission, 310
International Potash Agreement, 331
International Rubber Regulation Agreement, 326–29
International Sugar Conference, 329
International Timber Committee, 325
International tools of conservation policy, 304–46; when needed, 23
International Trade Organization, 320
International Union for the Protection of Nature, 313
International Zoölogical Congress, 312
Interstate Commerce Commission, 188, 353
Interstate coördination, 374–76
Industrialization, and non-Malthusian balance, 11, 14, 341–46; of underdeveloped countries, 340–46
Intertemporal distribution, *see* Time distribution
Investment, and costs in income tax, 175; and disinvestment, 103; anticyclical, 212–16; in foreign resources, 337, 340–42; public, 212–16; *vs.* conservation, 58
Irrationality in conservation decisions, 83
Irreversibility, 39, 251–53, 256–57, 265
Irrigation districts, 290

Jessup, Philip C., 307
Joint-production approach, 78, 79, 233, 380–81
Judiciary, function, 224

Kaldor, N., 184
Keller, Albert G., 140
Kiessling, O. E., 210
Knight, Frank H., 111
Knorr, K. E., 327, 331
Kroeber, Alfred J., 252

Labor, costs and supply of, during depressions, 204, 207, 210; in colonial areas, 9, 207, 210; mobility of, 72–73; stake of, in conservation, 236
Labor-intensive investments, 214
Labor unions, attitude on in-training program for personnel from foreign countries, 345
Ladangs, 328
Land Tenure Act, 154
Latin America, international conservation in, 312, 317–18; policy in, 25, 272
Law of diminishing returns, 205
League of Nations, 325, 329
Leases, 117, 152, 154, 157
Legal limits, of regulation, 299–301; of zoning, 283–86
Legislative branch of government, 224, 351–53, 365–69
Legislative Reference Service, 367–68
Legislative Reorganization Act, 366, 367
Leonard, Larry L., 308
Liefmann, Robert, 331
Limits, economic and legal, of regulation, 299–301; of zoning, 283–86
Liquidity, 117, 119
Loan contracts, 161, 162–64, 167
Loans, flexible payments on, 108, 162–65; markets for, 101, 102–3, 165, 202, 205; rationing, 103, 167
Logan, Leonard, 143
Logging practices, 295
London Convention, 312
Lorimer, Frank, 277
Losses, immoderate, in private economics, 87, 112, 117; in social economics, 251–53
Lump-sum taxes, 187
Lumpiness of collective goods, 243
Lumpy costs, 71, 204, 210

Machinery, mining, 133; rental of, 157
Macmahon, Arthur W., 365
Maine, Sir Henry James, 224
Maintenance, of mines, 209, 211
Malthusian checks, 8, 12, 16, 342; *see also* Neo-Malthusian doctrine
Man, as a geomorphologic agent, 3
Man–land ratio, 328
Manufacture, technological changes in, 6
Marginal cost, concept, 233; curves, 204–5; functions, 231, 241, 243, 380
Marginal costs and revenues, 69, 130, 204
Marginal revenue functions, 231, 241, 243, 380
Marginal utility of money, 104, 105, 207, 243
Marine resources, 5, 306–10, 313

Market, perfect, definition, 100
Market forms, 188–98; effect on use-rate relations, 68; in social economics, 234
Market interest rates, 100
Markets for assets and loans, 101–3, 155, 157, 166, 189, 195, 202, 205
Marsh, George P., 4
Marshall, Alfred, 232
Material-intensive investments, 214
Mathematical note on theory of conservation, 379–81
Maximization theory, 77–79, 232–34
Maximum cumulative or sustained use, 60
Maximum use rate, safe, 258
McFadyean, Sir Andrew, 327, 328
Merica, P. D., 336
Mesabi Range, 182, 338
Metal ores, taxes, 176
Migratory Bird Commission, 311
Migratory birds, *see* Birds
Military considerations, 7, 335–40
Mineral Leasing Act, 297, 298
Mineral production, regulation of, 297–99
Mineral resources and power department, 359, 364
Minerals, taxes on, 176
Mines, abandoned, deterioration of, 37, 74, 209, 211; submarginal, 209, 211, 218, 331, 332
Mining, adjustments to economic instability in, 209–12; discovery of new deposits in, 324; hydraulic (placer), 264–65, 298; relations through costs and revenues in, 74; strip, 264–65, 298; technological changes in, 6
Missouri Basin Committee, 370
Money veil, 54, 70
Monocultures, 124, 208
Monopolies, national, 306, 307, 308, 335–36
Monopolistic (and monopsonistic) conditions, 69, 102, 107, 166, 189–98, 234–35
Morgan, James N., 191
Morgenstern, Oskar, 89
Multiple-peak situation, 81

National Committee on Inheritance Taxation, 185
National Conservation Commission, 314
National interests, conflicting, 335
National monopolies, 306, 307, 308, 335–36
National Planning Commission, 355
National Recovery Act, 217
National Resources Planning Board, 354, 359; *see also* U.S. National Resources Board
National Security Resources Board, 358
Natural changes of flow, 266–67
Natural resources, focus on, 29
Necessary conditions for optimum state, equations for, 381
Neighborhood relations, 236, 282
Neo-Malthusian doctrine, 12, 29
Net revenues, maximizing, 77–79, 232–34; meaning, 105; taxes on, 174–77, 178; *see also* Revenues
Neumann, Johann von, 89
Neutrality, in resource relations, 44; in use-rate relations, 68
New York forest preserves, 48
Non-Malthusian equilibrium, 8–11, 341–42
Nonrenewable resources, *see* Stock resources
Nonuse, 48
North American Conservation Congress, 314
North Sea Convention, 307
Northwest States Development Association, 373, 374
Nuisance-abatement ordinances, 149, 299

Objectives, of conservation policy, 230–50; of this study, 19–26
Oklahoma Bankers Association, 167
Oil, a fugitive resource, 142, 143; discovery of new deposits, 324; in Venezuela, 341; taxes on, 176
Oil industry, relations through costs in, 74
Oil pools, potential exhaustion of, 336; unitized, 143, 197, 297
Oil production, regulation of, 297
Oligopoly, 69, 190–98, 235
Opportunity costs, 72
Optimum state of conservation, 58; distortions of, 246; *ex ante* and *ex post*, 77, 83, 89, 91, 93; mathematical equation for, 380–81; practical approximations to, 78, 80–83, 249–50; private, 21, 76–93, 181; private *vs.* social, 234;

social, 22, 230–50; theoretical determination of, 77–79, 232–34
Ore treatment, modern, 6
Oregon Forest Conservation Act, 293
Osborn, Fairfield, 12
Osborn, Frederick, 277
Osgood, W. H., 144
Outputs, deferring, 56
Overexpansion, *see* Speculative expansion
Ownership, 150; encouraging, 158; *see also* Property rights

Pacific Coast Board of Intergovernmental Relations, 371, 373
Pacific Coast Federal Regional Council, 358, 371
Pan-American Convention, 312, 313, 316
Pan American Scientific congresses, 317
Pan-American Union, 317–18
Pareto, Vilfredo, 247
Paris Convention, 310, 311, 314
Parks, Roland D., 178
Patent agreements, 322
Payments in kind, 164, 207
Pelzer, Karl J., 328
Penalties, for "depleting" crops, 135, 280
Pennsylvania Kohler Act, 298, 299
Perfect knowledge, *see* Best knowledge
Perfect market, definition, 100
Period, concept of, 31; planning, *see* Planning period
Periods of expansion and stagnation, 200, 202–19
Phillips, J. C., 311
Phosphates, regulation of use of, 298
Pigou, A. C., 191
Pinchot, Gifford, 15, 27, 314, 316, 317
Placer mining, 264–65, 298
Placerita Canyon, 142, 143
Planning, need for, 66–75
Planning agents, 28; uncertainty preferences of, 112; *see also* Conservation decisions
Planning and reviewing agency, 353–55, 359
Planning interval, defined, 31, 32; in agriculture and forestry, 80
Planning period, concept, 32; determining, 78, 79–80; effect of interest on, 99; effect of uncertainty on, 125; given, 79
Plans, *see* Conservation decisions; Planning agent; Planning period; Production plans
Plant associations, 257, 258
Plant species, 256, 258
Plantation agriculture, 328, 330, 342
Polak, J. J., 247
Polhamus, Loren G., 327
Police power, 272–73, 284, 285, 299
Policy, definition, 223; objectives and tools, 225; *vs.* social institution, 225; *see also* Conservation policy
Political obstacles to consolidation, 360
Pollution, of air and water, 258, 299
Pooling, 113, 120–22, 123–24
Population, and income, farm, 277; growth of ("explosion"), 7–11; pressure of, 342–43
"Pork barrel," 360
Positivism, 235, 245
Potash, 37, 209, 210, 331
Precipitation, 40, 43
President's Water Resources Commission, 347, 348, 353
Pressure groups, 225
Price agreements, 322; *see also* Commodity agreements
Price instability, and commodity agreements, 322–34
Price supports, 135, 136–37
Prices, 129–38; regressive, 106; under various market forms, 69, 189
Private economics, economic forces in, 95–219; optimum state in, 65–93; problems of, 21
Private enterprise, and a safe minimum standard, 260
Private-enterprise economy, public action in, 65
Privileged class, 245
Probability distribution, of expectations, 86, 111
Processing, regulation of, 299
Production and Marketing Administration, 278, 356
Production cycles, 201
Production function, in planning, 81
Production plans, 66–75; flexible, 87, 113, 116–19, 126
Productive services, as a flow resource, 41; "conserving" and "depleting," 132; price changes in, 132; regressive prices in, 107
Products, "conserving" and "depleting,"

Products (*continued*)
133–35; prices of, 129–32, 133–38; *see also* Prices
Profit taxes, 174–77
Progression, in income-changing forces, 106; in taxes, 171–73
Propaganda, 249
Property, 139–49
Property rights, concept and structure, 140, 148; imbalance of, 147–49; indefiniteness of, 141–45; instability of, 145–46; rigidity of, 146; *see also* Credit; Taxation; Tenancy
Property taxes, 177–84
Public credit aid, 165, 166
Public domain, 142, 144, 217
Public lands, leasing, 158; public purchase of submarginal lands, 217
Public management, of resources, 23, 301, 362
Public opinion, 225
Public policy, *see* Conservation policy; Policy
Public revenues and costs, 236; *see also* Social revenues and costs
Pure competition, 69, 189–98, 234
Pure monopoly, 190–98, 235

Questionnaires, 241–44

Rainfall, *see* Precipitation
Range, critical zone in, 257; *see also* Grazing; Grazing lands
Rate of flow, decreases in, 39
Rate of use, *see* Use rates
Rates, *see* Interest rates; Time-preference rates
"Reasonable, beneficial use," 287–90
Recovery costs, 71, 206
Recreation, and modern technology, 5
Recreational resources, critical zone in, 257; need for, 263
Refining, technological changes in, 6
Regions, federal, 372–74
Regression, in charges and prices, 106, 155, 163, 195; in taxes, 171
Regulation of practices, 287–303; why needed, 273
Relations, *see* Resources; Time; Use rates
Relocation assistance, 286
Renewable resources, *see* Flow resources
Rents, cash and share, 153, 156; fixed *vs.* flexible, 108, 155; regressive, 107, 155; sliding-scale, 153, 156
Residential uses, zoning for, 282
Resource, defined, 28–29
Resource use, *see* Cumulative use; Use rates
Resource users, education of, 275, 281
Resources, accessible and inaccessible, 336–40; changes in, 34; classification of, 29, 33, 35, 36, 38, 40, 42, 43; composite of, measuring, 55; *ex ante* and *ex post*, 30, 32; exhaustible and inexhaustible, 33; flow and stock, 35–53, 46–47; fugitive, 75, 141–45; measurement of, 35, 54; natural, human, and cultural, 29; "physical" and "economic," 29; primary and secondary, 43; relations between, 43–45, 262–65, 349–51, 381; submarginal, 29; *see also* Depletion; Flow Resources; Stock Resources
Revenue functions, marginal social, 231, 241, 243
Revenues, and exhaustibility, 33; competitiveness and complementarity through, 67, 74, 133, 194, 381; factors affecting changes of, 34; private and social, 231–44; redistributing, 98, 115, 178
Reversibility, of depletion, 39, 42–43
Richards, A. I., 90
Riesenfeld, Stefan A., 307
Risk, *vs.* uncertainty, 111
Risk premium, 116
River and Harbor Act, 369
River-basin authorities, 215, 357, 359, 370, 376–78
Robinson, Joan, 190, 232
Rogers, Lindsay, 365
Roosevelt, Franklin D., 276, 354
Roosevelt, Theodore, 15, 276, 314, 315, 316
Rousch, Gar A., 314
Rubber, 326–29

Safe minimum standard, 251–67; and direct tools, 302
Sales promotion, 191, 192, 195
Salmon, 310
Sanctions against aggressor nations, 336
"Sanitary revolution," 8
Sargant, C. H., 181
Scenic resources, zoning for, 282
Schloemer, F. C., 329

Schurz, Carl, 15
Scrap, 34, 68, 194, 211
Seals, fur, 308
Second cross partial derivatives, 381
Sectionalizing of investments, anticyclical, 213
Security considerations, 7, 335–40
Semantics, conservation as a field for, 19
Senate committees, 366–68
Severance taxes, 186
Shepard, Ward, 12
Sherman Act, 188, 198
Shriver, Harry C., 285
Simplifications of economic theory needed, 21; *see also* Approximations
Simpson, H. D., 183
Simultaneous economics, *see* Instantaneous economics
Site values, 41
Smart, William, 181
Smith, W. H., Jr., 153
Social costs, 231–38
Social economics, criteria and objectives, 221–67; problems of, 22; tools, 269–346
Social institutions, and conservation policy, 22, 225–29; and economic forces, 139; and flow resources, 46, 47; changes of, 12–14, 323–24; concept and structure of, 140, 225, 228; fields of, 140, 227; in underdeveloped countries, 343–46; *see also* Customs; Economic forces; Habit patterns
Social net revenues, maximizing, 232–34
Social optimum, *see* Optimum state of conservation
Social revenues, 231–44
Social sciences, and conservation, 26, 28, 30
Social weighting, 224, 244–46, 248
Soil, a composite of many resources, 39, 55; critical zone in, 256–57; regulation of use of, 290–93
Soil conservation, and price supports, 133–37; and strip or placer mining, 264–65, 298; education in, 279–81; effect of depressions on, 208; safe minimum standard in, 258, 259; zoning for, 282
Soil Conservation and Domestic Allotment Act, 280
Soil-conservation districts, 291–93
Soil Conservation Service, 278–80, 291–93, 331, 356
"Soil-depleting" crops, 133–35, 280
Soil depletion, 256–57; obscured, 70, 276–77
Soil erosion, in plantation agriculture, 328
Sombart, Werner, 124, 227
Spanish edition, 25, 272
Spaulding, E. Wilder, 315
Specialization in development, 252
Speculative expansion of crop production, 122, 123, 173, 208, 217, 322, 330
Spencer, Aubrey J., 155
Spreading, 113, 122–24
Stagnation, economic, 200, 202–19; of a social group, 252
Stand-by capacity, 335, 338
Starch, Elmer, 164
State compacts, 373, 374, 376
State governments, coördination in, 352, 359, 368–69; responsibilities of, 352; *see also* Federal regions; Federal-state coördination; Interstate coördination
State of conservation, 53–54; of depletion, 53; *see also* Optimum state
"State of the arts," 56
Static state of use, 49
Stationary state, and interest, 100; and uncertainty, 126; for last planning interval, 80
Step-by-step approximation of optimum state, 80, 82, 249
Stock resources, accessible and inaccessible, 336–40; adjustments to economic fluctuations in, 209–12; classes of, 36, 42; commodity agreements in, 331–34; conflicts of national interests in, 335; consolidating agencies for, 350–51, 353, 359–63, 364–65; defined, 35; depletion of, 4, 6; foreign, 337, 340–42; market forms in, 69; markets for assets in, 101; national monopolies in, 306, 307, 308, 335–36; renting, 154; taxes on, 171, 176; technological changes in, 46; world organizations needed for, 320–21
Stockpiles, military, 335, 337–40
Storage, of wheat, 173; reducing, 98
Strategic resources, 7, 335–40
Strip mining, 264–65, 298
Subsidies, 107, 122–24, 137, 238, 280; (compensation), 285, 286, 300; export, 330, 337; for depletion, 337; for exploration and research, 338

Substitution in resources, 6, 336, 339–40
Sugar, 329–31
Sulfur, regulation of use of, 298; taxes, 176
Sumner, William G., 140
Sunk costs, 71, 72–75, 206, 210; reducing, 98, 118
Supply, complementarity and competitiveness in, 44, 262, 264–65
Supply curves, and economic instability, 203–7; and monopoly, 191
Supply functions, community, 231
Sustained yield, 49; maximum, 60

Taconite, 182
Tariff wars, 330
Tax base, 171
Tax-saving certificates, 172
Taxation, 168–87; and property rights, 141
Taxes, 168–87; fixed *vs.* flexible, 108; progressive and regressive, 107, 171
Taxing power, in soil-conservation districts, 292
Taylor Grazing Act, 142, 144, 159, 217, 296, 300
Tea, 325
Teacher training in conservation, 277
Technological changes, 4–7, 34; and economic instability, 323–24; in stock resources, 46, 47
Technology, as a solution to resource problems, 16; of conservation, 17
Telescoping of investments, 213
Tenancy, 150–60; and property rights, 141; insecure, 145; *see also* Leases; Property rights; Rents
Tennessee Valley Authority, 279, 359, 376–77
Tenure, definition, 141, 150; indefinite, 141–45; insecure, 145, 152, 162, 197
Terminology, consistent, as objective, 19
Terracing, 82
Texas, oil pools in, 143; grazing regulation in, 296
Thair, Philip J., 172
Theory of conservation, mathematics of, 379–81; *see also* Maximization theory
Tiffany, H. T., 150
Timber, commodity agreements in, 324–26
Time, "period" and "instant" of, 31; relations to prices, 129–34; relations to taxes, 169–71; relations to uncertainty, 113, 114; *see also* Planning interval; Planning period
Time discount, 79, 98, 104, 116
Time distribution, of inputs and outputs, 56; of price changes, 129–34; of revenues, 98, 115; of use rates, 17, 21, 31, 51, 53, 98, 113, 114, 129–34, 169–71, 178–81
Time economics, 18, 20, 194–97, 206–8, 233–34
Time preference, 103–109; and income levels, 104, 247; in underdeveloped countries, 104, 207
Tin, 37, 210, 331, 332, 341
Tomasevich, Jozo, 308
Tools of conservation policy, 22–24, 216–19, 228–29; domestic, 271–303; international, 304–46; selecting, 237, 271; *see also* Economic forces; Education; Penalties; Regulation; Subsidies; Zoning
Toynbee, Arnold J., 252
Transportation improvements, 5, 12
Treaties and conventions, 145, 305–20, 325, 326
Troxel, Emery, 234
Truman Committee, 338

Uncertainty, 111–28; and conservation policy, 250; and economic instability, 202; and flexibility, 87, 113, 116; and monopolistic conditions, 195; and optimum state, 86, 250; ways of allowing for, 87, 112, 115–23
Uncertainty allowance, 87, 112, 115–23, 202; and the planning period, 79, 125; for political instability, 146; for price supports, 136; reducing, 115, 120, 123, 125, 155
Uncertainty bearing, by the public, 123
Uncertainty preference, 111, 114, 115
Underdeveloped countries, conservation education in, 343–46; depletion in, 9–10, 109; economic forces in, 108, 110; population explosion in, 9–11, 342; resource development in, 341–46; time-preference rates in, 104
United Nations Economic and Social Council, 334
United Nations F.A.O. and I.T.O., 319–20
United Nations Scientific Conference on Conservation and Utilization of Resources, 319

United States, applied portions of study confined to, 25, 271–72, 347
U.S. Bureau of Home Economics, 247
U.S. Bureau of Reclamation, 79, 164, 357, 360, 363
U.S. Bureau of the Budget, 354–55
U.S. Commission on Organization of the Executive Branch of the Government, 347, 358; *see also* Hoover Commission
U.S. Congress, *see* Congress
U.S. Constitution, 351–52
U.S. Corps of Engineers, 79, 357, 360, 363
U.S. Department of Agriculture, 135, 347, 353, 356, 357, 360, 361, 362, 363–64; *see also* specific agencies and services
U.S. Department of Commerce, 361
U.S. Department of Defense, 360, 362; *see also* U.S. Corps of Engineers
U.S. Department of State, 12, 315, 318, 333
U.S. Department of the Interior, 240, 347, 357, 360, 361, 362; *see also* specific agencies and bureaus
U.S. Fish and Wildlife Service, 301, 357
U.S. Forest Service, 159, 160, 279, 301, 356, 362
U.S. Grazing Service, 159, 279, 301
U.S. National Resources Board, 246; *see also* National Resources Planning Board
U.S. National Resources Committee, 144, 246, 277
U.S. Park Service, 301, 357
U.S. Supreme Court, 280, 285, 299, 353
Unitized oil pools, 143, 196, 297
Uranium ore, 299
Use, cumulative, *see* Cumulative use
Use rates, concept, 31; intertemporal distribution of, 17, 21, 51, 53; mathematical definition, 31, 379; maximum, 258; redistributing, 98, 170; relations between, 66–75, 169, 194, 381
Utilité, 232

Valuation of extramarket goods, 85–86, 239–48
Variable costs, 71, 204
Variable-payment plans, 156, 163
Variables, holding constant, 29; selection of, 21
Virgin resources, depletion of, 9
Vogt, William, 12, 318
Voters, education of, 274, 278

Wage rates, *see* Labor, costs
Wallace, Benjamin B., 331
Wants, changes of, 34
Warp, George A., 284
Wars, drain on resources, 7; *see also* Military considerations
Waste, of gas and oil, 297; of water, 290; *vs.* depletion, 57
Water, pollution of, 258, 298, 299; "reasonable, beneficial use of," 287–90; regulation of use of, 287–90; *see also* Ground water
Water resources, development, 377–78; safe minimum standards in, 258, 259
Watersheds, 377–78; zoning for, 282, 284
Watkins, M. W., 144, 348
Weber, Max, 227
Weeks Law, 217
Weighting, of distance in time, 52; social, 245
Wertheimer, Ralph B., 284
West Virginia Soil Conservation District Law, 198
Western world, capitalistic revolution in, 13; population explosion, 7–11
Whales, 5, 308
Wheat, 124, 173, 208, 325
Wicksell, Knut, 212
Wilderness areas, zoning to protect, 282
Wildlife, 141, 143, 145, 217, 305–14; regulation of use of, 296
Williams, C. C., Jr., 298
Wind erosion, 267, 291
"Wise use," 19, 49
Works Progress Administration (W.P.A.), 236
World organizations for conservation, 319–21

Yates, P. L., 331
Yield taxes, 186–87
Yount, H. W., 183

Zimmermann, Erich W., 33, 316
Zoning, 23, 282–86, 302

www.ingramcontent.com/pod-product-compliance
Lightning Source LLC
LaVergne TN
LVHW100250210926
848480LV00003B/879